IN THE CHARCUTERIE

IN THE
CHARCUTERIE

How to Make Sausage, Salumi, Pâtés, Roasts,
Confits, and Other Meaty Goods

TAYLOR BOETTICHER and TOPONIA MILLER of The Fatted Calf
Photography by Alex Farnum

◨ SQUARE PEG

CONTENTS

INTRODUCTION: COME ON IN

WHEN YOU FIRST walk through the doors of the charcuterie, it feels as if you've entered an enchanted world of meaty wonders. The aroma of crispy-skinned pork roast fills the shop, inviting you to try a bite. Our cases are filled with pâtés, *salumi*, sausages, roasts, and terrines – and when the meat counter crew offers you a slice of the fennel-flecked *sbriciolona* and a piece of head cheese, it's hard to say no. Walk back into our kitchen and you'll smell spices toasting, bones roasting, and broths simmering. Someone is churning out coils of fresh sausage from the hand-cranked stuffer, and someone else is hanging huge, freshly-cased *cotechino* on hooks for fermentation. We're hand-shredding a veritable vat of duck rillettes, seasoning it with freshly chopped thyme, then packing it into jars and sealing each with a creamy layer of duck fat. Bacon has just finished in the smoker! Go ahead and tear a hot piece off the end of the glistening slab. Peer into our curing room where row upon row of salami,

guanciale, and pancetta hang quietly, patiently, enrobed in a delicate snowy bloom of mould.

If you are curious and want to know how all of this works – if a hunger stirs inside of you and you feel somehow strangely at home – then you are in the right place. Welcome to the Fatted Calf Charcuterie.

A charcuterie is a bit of a strange business, and the Fatted Calf is an unusual charcuterie. More than just a butcher shop, we offer meaty goods and services that are varied and unique. Pick out a duck or a slab of ribs and we'll cook it up for you. Want us to wrap and season your pork tenderloin? No problem. Craving a dish you ate at a little roadside restaurant in Burgundy and need wads of lacy caul fat and pork jowl to recreate it? We have it here. At the Fatted Calf, we butcher whole hogs, goats, and lamb. We sell chicken, duck, rabbit, and quail. We make our own sausage, pâtés, terrines, and potted meats. We cure *salumi*, smoke pastrami, and roast porchetta. We think about, talk about, and share our love of good meat from the moment we open until we close each day. We also teach butchery and charcuterie-making classes because we want to pass on the knowledge we've acquired to you.

Most people associate 'charcuterie' with a trip to Paris or a delicious platter of meats at a restaurant. But real charcuterie goes well beyond that. The charcutier transforms the bounty of the farm and forest into a delicious subset of cuisine, which ranges from sausages and hams to stuffed game birds and elaborate roasts. At its most basic level, charcuterie is the technique of seasoning, processing, and preserving meat. But it is also a way of preserving food cultures and traditions, and enriching our daily habit of breaking bread. It is a holistic approach to cooking and eating meat and a rewarding, hands-on way to connect with our food. At the Fatted Calf, charcuterie is a way of life – an approach to cooking and eating that celebrates the pleasures of the table. 'Charcuterie' can be a succulent confit duck leg atop a bed of crisp greens; a rich and meaty stew on a cold winter's night; or a picnic blanket spread in the shade of an old tree, laden with half-empty crocks of pâté, dishes with pickled vegetables, and slices of fragrant salami.

For as long as people have needed to preserve their meat, charcuterie has existed in one form or another. However, it was the Romans (sticklers for rules that they were) who first codified the laws of meat preservation. Charcuterie began to blossom in earnest in France during the Middle Ages, when an official guild system was put in place to regulate the production of processed meat products. This gave birth to the delicatessen-like shops, also known as *charcuteries*, where the products were sold.

Over time, many of the techniques first developed in France and Italy (such as salting hams for prosciutti and processing meat into loaves or terrines) spread to the neighbouring countries of Germany, Spain, and beyond. Local predilections and ingredient availability produced countless regional charcuterie specialities, including speck, *Jamón Serrano*, and many, many more. In both Europe and the Americas, the industrial revolution hastened the charcuterie boom: swarms of people left the farm to work in urban factories, and thus lost access to proper kitchens and fresh meat products. The charcuterie or delicatessen became a necessity, a place where you could buy the makings of a simple meal with minimal effort. By the turn of the twentieth century, in major cities like Milan, Paris, or New York, you couldn't throw a stone without hitting a *salumeria*, charcuterie, or deli.

Then along came suburbs, the supermarket, and the rise of industrial agriculture. Production moved away from small, local shops and into bigger and bigger meat processing plants. Decades- or centuries-old recipes were dumbed down for efficiency's sake, quality was sacrificed for quantity, and many regional specialities were lost or forgotten. Meats came pre-sliced, prepackaged, and loaded with unhealthy preservatives. Even in Europe, charcuterie's birthplace, the traditions started fading. By the time we visited in the 1990s and early 2000s, there were hardly any young people learning the trade; charcuteries and *salumerias* were run by a handful of people from the older generation. An empire with no heir.

When we started the Fatted Calf in a sublet kitchen in San Francisco's Dogpatch neighbourhood in 2003, we had only an inkling of what we wanted to achieve. What we did know we had gleaned from dated texts and from our meat mentors in professional kitchens and butcher shops. We knew that we wanted to honour the traditions of the craft but infuse them with our own quirky sensibilities. We borrowed from the old ways but weren't shy about incorporating the new. We sourced the best meat we could find, always from small family farms that used humane and sustainable methods. We used locally grown produce and foraged mushrooms, and the products we made and sold followed a rhythm of seasonality.

We were also staying up late, linking sausage and loading terrines well into the evening. We did a lot of dishes, cursed our beat-up old machinery, and suffered a few heartbreaking disappointments. In the beginning, we were like meat gypsies, hawking our wares out of coolers at Bay Area farmers' markets, huddled under our tent while gusts of wind sent freezing January rain on to our neatly arranged salami and terrines. But in spite of the weather, people came and tried our wild mushroom terrine, Toulouse sausage, and saucisson. And then the following week, they came back. They told their friends about us, and the Fatted Calf grew. When our little kitchen in San Francisco could no longer contain us, we moved to new digs in Napa. A few years later, we opened a second location in San Francisco's Hayes Valley neighbourhood.

Today the craft of charcuterie is experiencing a revival. While it started with chefs and a handful of artisanal producers, now it has practically gone mainstream thanks to an enlightened market of consumers who are curious about the origins of their food and hip to the fact that industrial agriculture and mass-produced meat products are not good for the planet or the population. We are proud to be a part of this movement and we want you to join us.

This book has something for everyone, whether you're a sceptical ex-vegan, scimitar-wielding novice, or seasoned old pro. When you walk into a butcher's shop and spy a pork shoulder in the case, we want you to see more than just a hunk of meat. We want you to see all of the possibilities that pork shoulder has to offer, from shoulder chops and stuffed roasts to picnic hams and salami. We want you, knife in hand, to experience what it is like to break a whole animal into its parts. We want to share with you not only the knowledge of butchering and cooking

we have accumulated through our work, but also the respect we have for the raw ingredients, the satisfaction we derive from working in the kitchen, and the pleasure of sitting at the table with friends and family to eat what you have created. We want people to better understand the processes of charcuterie by participating in it. So we invite you to slip into our greasy clogs for just a little while. In this book, we'll ask you to plunge your hands into a freshly ground forcemeat to make sausage, inhale the intense perfume of a spice blend, confidently carve a roast, and more. And at the end, you get to enjoy the delicious results of your labour and passion.

We cook a lot, not just in the charcuterie but at home as well. The methods and recipes in this book are based on our professional experience of working in a charcuterie for roughly a decade – but they are also written with the home cook in mind. Quite a few of the recipes and methods presented here are simple to master, and we hope that they'll edge their way into your culinary repertoire with ease. Others are more challenging, multi-stepped processes that require several days or even weeks.

Charcuterie is a discipline that requires patience. Allowing plenty of time and space is the key to successful smoking, curing, and terrine- or sausage-making. The gratification is far from immediate, and may seem out of sync with our modern way of life. But we believe that there is a place for these meaty meditations: they can teach us truths about history, community, sustainability, and self-sufficiency. With *In the Charcuterie*, we want you to take the same pleasure from butchering, cooking, and preserving your meat as you do savouring it at the table.

White Pepper

MACE WHOLE

NUTMEG.

BACON Rub — NEEDS Aleppo

CARAWAY

Esplette

RL GRAY TEA

pancetta SKIN ON

FENUGREEK

a Cure 4-3

Beef Pastrami RU

Fleur sel

a Cure 5/7

-6/2

6-5

da Cure 4-3

GREEN P

THE CHARCUTIER'S
PANTRY

On the shelves of the Fatted Calf Charcuterie, you'll find buckets brimming with salt and stacked containers crammed with whole spices. Tubs of garlic, onions, and shallots are stored underneath trays of drying lavender, thyme, and oregano. Baskets of chanterelles and bins of herbs and citrus are kept in the cooler. Stashes of dried apricots and porcini sit alongside jars of dried *chile de árbol* and cayenne peppers. Meat makes up the core of the charcuterie, but our pantry provides us with a palette of flavours with which to work.

Stocking the charcutier's pantry is fairly simple. Most of the items you will need are commonplace or easily obtainable. Many may already be in your pantry. But the attention to detail paid to these staples is essential. Knowing which salts to use for each purpose is crucial. Understanding how to extract the most flavour from your spices is key. Learning to prepare *herbes de Provence*, mushroom *duxelles*, and brandied fruits adds breadth to your repertoire. Your well-provisioned pantry will endow you with the flexibility to take a simple pork shoulder and turn it into a succulent, seasoned roast, Toulouse-style sausages, spicy *ciccioli*, *cacciatorini* salami, or dozens of other preparations you'll find in these pages.

Salt

At its most basic, charcuterie can be reduced to two elements: meat and salt, in an endless variety of combinations and methods, and with a changing cast of supporting characters. Salt is both seasoning and preservative. Different types of sea salts and curing salts serve specific purposes in the charcuterie, and discerning how each affects the taste, texture, and longevity of meat is a valuable skill.

Seasoning with Salt

A sprinkle of salt enhances the taste of food by triggering receptors in our palates to allow us to perceive saltiness as well as subtle nuances of flavour. When practised with thoughtfulness and confidence, salting creates foods with fuller, sharper flavours, and thus dramatically increases the pleasure of dining.

Unrefined sea salt is the most natural type of salt available and is the only type we use in the charcuterie for seasoning. Table salts contain additives or other substances to make them pour more easily or to increase their nutritional value. These additives, such as iodine, fluoride, and aluminium,

even in minute quantities, can cause an unpleasant chemical taste. A variety of good-quality sea salts is available in a range of colours and textures, each with its own unique minerality, a reflection of its origins and the techniques used to harvest it. The three types that we always have on hand in our pantry are fine sea salt, coarse sea salt, and *fleur de sel*.

We use **fine sea salt** more than any other for our everyday seasoning. Whether we are making a quick brine, rubbing a slab of ribs, preparing fresh sausage, salting duck legs for confit, preserving lemons, or pickling cucumbers, our go-to is a finely grained sea salt. It disperses faster and more evenly than larger-grained versions, making it more practical for most applications. Nearly all of the recipes in this book call for fine sea salt. If you are accustomed to using a typical table salt, which has a slightly smaller grain and contains a host of strongly flavoured additives, you may find you need a bit more sea salt. Even among fine sea salts, the flavour and salinity can vary.

Coarse sea salt, such as the chunky salt from Trapani, Sicily, is prised for its texture. The large granules are extremely slow to melt or blend with foods, even when exposed to heat, enabling them to impart not only an intense flavour but also a crunchy texture to the crust on a thick cut of meat or a roast. In the charcuterie we often rub coarse sea salt on skin-on pork roasts, such as The Cuban on page 181, to produce irresistibly crispy crackling.

Fleur de sel is the delicate hand-harvested salt collected from the top layer of salt pans (large ponds of briny seawater that dry due to natural evaporation, leaving behind only salt). Traditional French *fleur de sel* from the coastal waters of Guérande or the Camargue has a flaky, brittle texture and subtle flavour. (Maldon salt is another nice option.) *Fleur de sel* is generally considered a 'finishing' salt, added to foods just before serving, and should you find a cooked steak or chop in need of a little seasoning, *fleur de sel*

is a good choice. In the charcuterie, we prefer *fleur de sel* for seasoning foie gras.

When and How to Season

Throughout this book you will find numerous specific instructions on when to season, how to season, and how much to season. The question of how much and when is very important, especially for confits and cured meats, so pay attention to specific meat-to-salt ratios and time periods for marinating or brining. At other times seasoning is a little more subjective. In general, though, meat should be seasoned more liberally and for longer periods than other foods. The thicker the cut, the more salt you will use, and the longer the period of pre-seasoning required. Fattier cuts or meat on the bone should be seasoned more and for a longer time than leaner or boneless cuts.

Salting for Preservation

Salt in its many forms has been used to preserve foods for thousands of years. It works by creating a hostile environment for certain microorganisms: osmosis dehydrates bacterial cells, thereby inhibiting the bacterial growth that can cause spoilage. Sauerkraut (page 319) is a great example of this process. Sea salt is massaged into freshly cut cabbage, breaking down the cell walls and extracting moisture, creating a natural brine that can preserve the cabbage for months.

Different salts produce varying results of preservation. Although sea salt can be an effective preservative for many foods, its usefulness on its own as a meat preservative is limited. A light dusting of it might extend the shelf life of the meat for a few days, but not indefinitely. Large amounts can preserve meat for longer periods but can also make it unpalatable. A more effective preservative is sodium nitrate, a naturally occurring mineral that is present in many plants and organisms. Virtually everything that grows in the ground draws sodium nitrate from the soil,

Despite the fact that they are an extremely effective meat preservative, nitrate and the resulting nitrite have had a pretty bad rap in recent years, triggering a rash of nitrate-free hot dogs, salami, and bacon at the meat market. The nitrate scare began in the early 1970s with a much-publicised study that implicated a link between nitrates and cancer. Subsequent studies did not support the initial conclusion. Regardless, the notion that nitrates cause cancer lodged in the public's mind and the demand for nitrate-free products with a similar look and taste to those with nitrates prompted producers to seek an alternative. The most popular substitute for traditional sodium nitrate is celery juice, which, incidentally, contains an abundance of sodium nitrate that converts to sodium nitrite just as the traditional preservative does. But celery juice allows producers to claim that their products have 'no added nitrates', are 'organically cured', or are 'nitrate-free'. These claims are essentially fraudulent, as so-called naturally cured meats sometimes contain more than twice as much sodium nitrate as traditionally cured products.

including vegetables such as carrots and celery. The bacteria that cause spoilage need moisture, oxygen, and food to live. In addition to drawing out moisture as other salts do, sodium nitrate, which converts to sodium nitrite during the curing process, acts as an extremely effective antimicrobial and antioxidant. It thus prevents the growth of most harmful bacteria, including *Clostridium botulinum*, which can cause deadly botulism. There are two types of curing salt commonly used for preserving meats, which contain either sodium nitrite or a combination of sodium nitrate and nitrite. These are generally referred to as curing salt no. 1 and curing salt no. 2.

Curing salt no. 1, also known as Prague powder no. 1 and by various brand names, is pink (to save it from being accidentally mistaken for table salt) and is 6.25 per cent sodium nitrite and 93.75 per cent sodium chloride (salt). It is used to preserve cooked or smoked sausage, salami, and meats such as bacon and smoked ham.

Curing salt no. 2, also known as Prague powder no. 2 and by various brand names, contains 6.25 per cent sodium nitrite, a smaller percentage of sodium nitrate, and the balance is sodium chloride. Depending on the brand it can be either pink or white. It is specifically formulated for dry curing products such as salami, which won't be cooked or smoked. It works over a longer period of time than curing salt no. 1, with the sodium nitrate breaking down to sodium nitrite and then to nitric oxide, which sets the colour and ensures the meat is safely cured.

Spices

Spices provide a range and depth of flavour that can bring a sense of drama to your everyday cooking. Quality whole spices, from the ubiquitous black peppercorn to the enticingly perfumed star anise, are key components of the recipes in this book. Spice blends such as *quatre épices* (four spices) provide a certain *je ne sais quoi* to confits and rillettes. *Ras el hanout*, a complex North African blend, imparts glamour to ordinary lamb chops. Having a good selection of fresh spices in your charcuterie pantry is also essential. Every sausage, salami, and terrine derives its distinctive flavour from a carefully orchestrated combination of spices.

Purchasing and Storing Spices

Carefully source and store your spices to enjoy their maximum benefits. Although most spices are dried, their potency and aroma fade over time.

Some particularly oily spices, such as fennel and anise, can become bitter or rancid. To guarantee freshness, purchase spices in bulk, directly from a spice merchant if possible. Skip the spice aisle at the supermarket. Those spices are often stale and many contain additives such as wheat flour, which producers are not required to list on the label. Even if your area lacks a good spice merchant, many now sell online. Ethnic groceries with a high turnover of stock are a good alternative. Purchase whole spices whenever possible and in quantities that you can reasonably use within a year, to keep the selection in your pantry fresh. You might easily use half a kilo of black peppercorns within 12 months, but more modest quantities of nutmeg or aniseed, which should be purchased in smaller quantities.

CHARCUTIER'S ESSENTIAL SPICES

Aleppo pepper flakes (*Pul biber*)	Mustard seeds (yellow)
Allspice, whole	Nutmeg
Aniseed	Paprika, Hungarian
Cardamom	Paprika or *pimentón*
Cayenne peppers	– Spanish smoked; sweet, medium-hot, and hot varieties
Celery seeds	
Chilli flakes	– mild unsmoked; also known as 'Spanish paprika'
Cinnamon sticks	
Cloves	
Coriander seeds	Peppercorns
Cumin seeds	– black
Fennel pollen	– white
Fennel seeds	– Sichuan
Fenugreek	*Piment d'Espelette*
Ginger	*Ras el hanout*
Juniper berries	Star anise
Mace	Turmeric

Spice blend for Lamb and
Herb Meatballs, page 206

Heat, light, humidity, and exposure to air can all degrade the quality of your spices. Store them in airtight containers away from direct sunlight and the heat of your oven. If labelling the container with the purchase date seems like a bit much, follow this rule: if you cannot recall when you purchased a spice, it is probably time to discard it and start afresh.

Toasting and Grinding Spices

Most whole spices, even black peppercorns and cinnamon sticks, benefit from a light toasting to release their essential oils and intensify their flavours. Toasting can be done on the hob in a dry heavy skillet over a low heat or on a baking sheet in a 170°C (gas 3) oven. Toast the spices for roughly 5 minutes, until they give off a pungent aroma. Always allow spices to cool to room temperature before grinding.

You can grind the spices by hand using a pestle and mortar or in an electric spice or coffee grinder.

A pestle and mortar are ideal for grinding small amounts of spices or for producing ground spices with a coarser, more rustic texture. However, if you are grinding large quantities or need finely ground spices for a sausage or salami recipe, an electric grinder works best. It is wise to invest in a grinder that will be used only for spices, to avoid sipping a cup of cumin-scented coffee.

Growing and Foraging Spices

Spices often seem exotic, an import from a faraway land. Although many spices require a hot, tropical climate to thrive, there are plenty of spices that you can grow or forage in your own garden. Coriander, fenugreek, celery, and mustard yield both green, leafy herbs and dried seeds. Growing and drying your own chillies can easily provide you with an annual supply of cayenne and paprika. In many places, fennel pollen and fennel seeds can be wild harvested. When the yellow fennel flowers bloom, their pollen rests atop

FIVE SPICE

This recipe should rightly be called Six Spice, since six is the number of spices it contains (though it is not uncommon for this Asian spice blend to contain more or fewer than the five of its name). One spicy theory posits that the five refers not to the number of ingredients but to the number of tastes (sweet, sour, bitter, savoury, and salty) that you experience when you eat it. Whatever its etymology, this slightly exotic, powerfully alluring blend is a great match for pork, duck, and even fattier cuts of lamb. Keep a small jar on hand to spice up a marinade, rub on a rack of ribs, or season a braise or stew. **MAKES ABOUT 35G**

8 star anise pods
2 teaspoons Sichuan or black peppercorns
1 cinnamon stick
1 tablespoon fennel seeds
1/2 teaspoon whole cloves
1 dried cayenne pepper, stemmed but not seeded

Preheat the oven to 170°C (gas 3).

Spread the star anise, peppercorns, cinnamon stick, fennel seeds, and cloves on a baking sheet. Toast in the oven until fragrant, about 5 minutes. Allow to cool to room temperature.

In a spice grinder, combine the cooled spices and the chilli and grind until reduced to a fine powder. Transfer to a tightly capped jar or other airtight container and store in a cool, dark place. Use the blend within 3 months, as it will lose its potency over time.

their crowns and can be gently shaken into a bag. If you have an abundance of fennel flowers, cut some of them and hang them over a baking sheet to collect the pollen that drops as they dry. Pollinated flowers left on the plant will produce a wealth of fennel seeds that can be harvested and dried.

Herbs and Alliums

We keep a few horse troughs planted with sage, thyme, tarragon, chives, rosemary, and oregano outside the entrance to the Fatted Calf shop in Napa. Passers-by frequently pause and touch the feathery flowering tops of parsley, crush leaves of lemon thyme between their fingers, or pluck a leaf or two of delicate marjoram and inhale. Both fresh and dried herbs and alliums (garlic, onions, shallots, leeks, chives) are used in the Fatted Calf kitchen for a variety of effects. Sage and thyme are comforting and familiar. Aromatic blends, such as lavender-scented *herbes de Provence,* can transport you to a verdant hillside in France's sunny south. Pungent garlic and oregano add savoury depth. Herbs and alliums breathe life into a standard recipe and provide a nuanced counterpoint to rich, fatty flavours.

Fresh and Dried Herbs
Both fresh and dried herbs play an important role in the charcuterie. Fresh herbs in a marinade or sauce infuse meats with bright flavours. They provide an enticing garnish for fresh sausages and pâtés and add the vibrant flavours of the garden and the aroma of the woods and fields to your meat dishes. But dried herbs are sometimes favoured; they can be pulverised in a spice grinder and more easily dispersed throughout a sausage or terrine. In general, dry-cured meats such as salami call for dried herbs. Fresh herbs would not survive the drying and ageing process intact and would add moisture where it is unwanted.

We never purchase dried herbs, preferring to dry our own throughout the year, a practice that provides us with a constant supply of pungent, organic freshly dried ingredients. Drying herbs such as bay, oregano, rosemary, savory, sage, and thyme is extremely simple and well worth the effort. If you are harvesting your own, be sure to cut off the tops of the plants before they flower. If you are buying them, set aside the desired quantity for drying while the herbs are still fresh and sprightly. Remove any blemished or yellowed leaves. Set a cooling rack over a baking sheet and lay the sprigs in an uncrowded layer on the rack. Place in a dry, well-ventilated location away from direct sunlight for 10–14 days, until dry. When the herbs are crumbly and the stems snap readily when bent, the leaves are ready. Remove the leaves from the stems and store them in a tightly capped glass jar or other airtight container in a cool, dark, dry spot for up to 6 months.

Alliums
Garlic, onions, shallots, leeks, and green onions are vital to charcuterie. Most recipes are enhanced by their pungency. We use alliums in their fresh, or green-topped, state in the spring and dried throughout the year. In our wild, modern world, it is sometimes easy to forget that these pantry staples are a seasonal crop. New garlic tastes very different from sprouting garlic. There are also huge differences between varieties. Although the recipes in this book suggest specific amounts, trust your tongue. It is not uncommon for us to add a little more or use a little less depending on the potency of a particular allium. When purchasing fresh alliums, look for lively green tops and sturdy, blemish-free root ends. Dried alliums should be firm with no visible green sprouts (a clear indicator that they are over the hill).

Drying fennel pollen

HERBES DE PROVENCE

Herbes de Provence is the name given to a blend of herbs commonly used in the south of France. Which herbs are used can vary slightly, but they are always dried. In the charcuterie, we use this mixture in marinades, as a rub for grilled and roasted meats, and as a seasoning blend for sausage and salami.

Forget the fancy, overpriced jar at the gourmet shop. The best blend is the one you make yourself in small batches using your own dried herbs. It can be assembled in just a few minutes and keeps well for months. If you have an abundance of dried herbs, this blend makes a lovely gift.

MAKES ABOUT 25G

6 dried bay leaves, crumbled

2 tablespoons dried rosemary leaves

2 tablespoons dried thyme leaves

2 tablespoons dried oregano leaves

2 teaspoons crumbled dried lavender buds
(English or another culinary variety)

In a small bowl, combine the bay, rosemary, thyme, oregano, and lavender and mix well. Transfer to a tightly capped jar or other airtight container and store in a cool, dark place. Use the blend within 3–4 months, as it will lose its potency over time.

To use, pulverise the herbs in a pestle and mortar if you prefer a coarser texture, or in a spice grinder for a finely textured blend.

Other Indispensable Staples: Fruits, Roots, and Mushrooms

Stockpiles of dried fruits, vats of salted lemons; bins of fragrant mandarins; jars of brandied currants and of prunes, knobs of gnarly ginger; rough-hewn roots of pungent horseradish, baskets of earthy fresh mushrooms; canisters of dried porcini and morels stored after the close of their brief seasons – all have an important place in the charcutier's pantry. These indispensable staples add flavour – both bold and subtle – and texture to a range of meaty dishes. They give the charcutier the flexibility to knock out a quick marinade, rustle up a stuffing, or whisk together a simple sauce with what is to hand.

Fruits

All year long, the natural sweetness and acidity of fruits bring balance to salty, savoury meats in the form of stuffings, garnishes, or accompaniments. Winter is the time of citrus and dried fruits. The intensely perfumed zest of lemons and oranges brightens marinades, fresh and dried sausages, pâtés, and terrines. Thinly sliced mandarins and oily black olives find their way inside pork roasts. A variation on the traditional *salsa verde* made with preserved Meyer lemon and chilli flakes complements braised and roasted meat. Dried currants sweeten *crépinettes*. Plumped prunes adorn a duck terrine, and dried figs are poached in red wine or pickled for nibbling alongside salami.

PRESERVED MEYER LEMONS

In the winter, when Meyer lemons ripen, folks show up at the charcuterie with boxes and bags full of these fragrant beauties. Although the aroma of freshly picked Meyers is heavenly, the flavour can be strong – it has been described as reminiscent of furniture polish – so strong, in fact, that it can overpower even hearty meat dishes. But salt and time have the ability to tame both the flavour and texture of these lemons, rendering them supple and pleasingly aromatic.

Preserving Meyer lemons is not an exact science. This recipe will make as many lemons as you happen to have. Use as few or as many of the spices as you like, or omit them altogether for a straightforward version. The tender skin of the preserved lemons can be chopped and used in sauces, braises, stuffings, and as a garnish for *crépinettes*. The pulp is an excellent addition to marinades.

Meyer lemons
Fine sea salt
Coriander seeds
Cinnamon sticks
Whole cloves
Peppercorns
Dried bay leaves
Freshly squeezed lemon juice

Choose a glass or ceramic jar large enough to accommodate your lemons.

Cut each lemon lengthwise from the tip towards the stem end into quarters, pausing 2.5cm before the stem end, so that the quarters remain attached. Holding the lemon over a bowl to catch any juice and salt, open like a flower. Cover its interior with a thick coating of salt, then gently press the quarters back together.

Pour a thin layer of salt over the bottom of the jar or crock. Layer the salted lemons in the jar along with a few of each of the spices and some of the salt and juice that has accumulated in the bowl. Weight the finished lemons with a small plate and let sit overnight.

The following day, remove the weight. If the level of the brine has not risen above the topmost layer of lemons, add additional lemon juice until the lemons are immersed in liquid. Cover the jar and let it sit at room temperature for about 30 days before using. You can leave the lemons at room temperature for up to a few months, but refrigeration will extend their shelf life for up to 6 months.

To use, lightly rinse a lemon under cool running water to remove the excess salt. Scrape out the pulpy interior of the lemon with a spoon, then chop the tender skin. Use pulp and skin as directed above or in individual recipes.

DRIED FRUIT IN BRANDY

Dried fruit plumped in black tea and brandy adds a sophisticated sweetness to many meaty preparations. In the charcuterie we keep several types of fruit in brandy on hand to garnish *crépinettes* and terrines, stuff roasts, or serve alongside smoked or roasted meats. Prunes and currants are staples in our pantry, but you can use this same method to preserve dried cherries, figs, apricots, or pears. **MAKES ABOUT 1 LITRE**

600g pitted prunes, dried currants, or other dried fruit

2 tablespoons honey

500ml piping-hot freshly brewed strong black tea (such as Assam or Darjeeling)

240ml good-quality brandy, Cognac, or Armagnac

Pack the fruit into a clean 1-litre jar. Stir the honey into the hot tea and pour over the fruit. Top with the brandy. Cover and refrigerate for at least 2 days before using. The fruit keeps well refrigerated for several months.

The long cold spell is broken in spring with the arrival of cherries, the season's first stone fruit, which make an excellent fruit *mostarda*. The heat of late spring and of summer brings on more stone fruits – apricots, peaches, nectarines, plums – for chutneys. Late in the season, figs appear, a favourite for stuffing quail. As autumn nights grow cooler and longer, pears, quinces, and apples fill the pantry with their honeyed aroma. A pork roast filled with roasted apples, walnuts, and sage speaks of the season. So, too, do poached pears folded into a salad of bacon and bitter greens, and fragrant, astringent quinces stirred into a rich lamb stew. Each season's offerings are anticipated, their usefulness appreciated in the kitchen and their flavour savoured at the table.

Roots

There is no substitute for fresh ginger, horseradish, and turmeric for meaty marinades, sauces, or seasonings. Their oddly scruffy exteriors belie the powerful, piercing flavours contained in their flesh. Fresh roots should be firm and wrinkle-free. Purchase them in reasonable quantities to maintain a fresh supply. To use, cut away the portion of the root you intend to use, then peel away the skin and grate the flesh. Freshly grated ginger and turmeric and especially freshly grated horseradish (which becomes bitter quite quickly) should be used right away. Store each of the cut roots in its own sealed bag in the fridge.

Fresh Wild Mushrooms

We use fresh wild mushrooms in the charcuterie every chance we get. Their flavours echo the rich flavours of meat, and the two make a tasty pair. Links of wild mushroom sausage loaded with sautéed morels, *crépinettes* studded with chunks of browned porcini, broth scented with unctuous chanterelles, and a glorious pork roast stuffed with wild mushroom *duxelles* (page 189) are highly anticipated seasonal treats. Finding a good local source for fresh wild mushrooms is essential. Or better yet, get off your backside and have a look around your area. If you like to cook with mushrooms as much as we do, some homework and a walk in the woods will pay off. Local mycological societies can provide you with a wealth of information about foraging in your area and help you to network with experienced foragers. Always consult a field guide or two when foraging and only eat mushrooms that you are completely confident are edible.

Dried Wild Mushrooms

For sauces, compound butters, meaty braises, even for adding to a terrine, dried wild mushrooms provide earthy flavour any time. A good-quality dried mushroom is often preferable to a less than fabulous fresh mushroom. If you are substituting dried mushrooms for fresh in a recipe, you will need to use about 85g for every 450g called for in the recipe.

We prefer an overnight soak in cold water to rehydrate dried mushrooms rather than the fast soak in hot water often recommended. Cold water preserves the flavour of the mushrooms a bit better and the long soak allows for an even distribution of moisture. Cover the mushrooms with three to four times the volume of cold water and let them sit overnight. The following day, gently remove the mushrooms from their soaking liquid, pat them dry (do not squeeze) and prepare them as you would fresh mushrooms. If you plan to use the soaking liquid, strain it through a fine-mesh sieve to remove any dirt or sand that might have been clinging to the mushrooms.

Truffles

Truffles are not quite a mushroom but the related fruiting body of the ascomycetous fungi, born of the precarious relationship between clement weather, rich, chalky soil, and an underground network of mycelia nurtured in the haven of particular tree roots. The resulting knobby tuber with its beguiling, musky aroma is the subject of much gastronomic waxing.

It is the black or Périgord truffle (*Tuber melanosporum*), named for the region in France where it is often foraged, that is particularly useful in upping the charcuterie ante. A few well-placed shavings of fresh black truffle add a swoon factor to pâtés, mousses, fresh sausage, salami, and other preparations. A host of truffle products are available at the market, from oils and pastes to salts and vodkas, and most of them are not good. Many of them don't even contain actual truffle. If you want the flavour of truffle, save your pennies for the real deal. Purchase fresh truffles during the season, which begins in the last days of autumn and stretches well into winter. Take a good whiff before you sign on the dotted line. Fresh black truffles should have a powerful aroma. Truffles are perishable and need to be used immediately or carefully stored in an airtight container lest their ephemeral aroma dissipate. In the charcuterie, we store our fresh truffles loosely swaddled in cheesecloth and placed in roomy covered jars filled partway with Arborio rice and fresh eggs, which benefit from the aroma of the truffles. A little truffle goes a long way. Shave thin slices with a truffle shaver or small mandolin, or grate the truffles with a rasp-type grater.

THE CHARCUTIER'S WILD MUSHROOM DUXELLES

Stuffed into a roast, folded into a forcemeat for sausage or pâté, or used as a base for a decadent mushroom sauce, this robust *duxelles*, the classic French preparation of chopped sautéed mushrooms, is nearly always on hand in our pantry when wild mushrooms are abundant. In the charcuterie, we prepare *duxelles* using lard or duck fat (of course!) and finish it with *gelée* (see page 52) which both seasons the mushrooms and binds the mixture, important if you plan to use it as a stuffing for meats.

For this preparation, fresh wild mushrooms work best. Morels are a favourite but porcini, chanterelles, black trumpets, hedgehogs, or a mix of whatever wild mushrooms you can find will do. **MAKES ABOUT 650G**

2 tablespoons lard or rendered duck fat
450g wild mushrooms, trimmed and sliced
Fine sea salt
40g minced shallot
3 tablespoons chopped fresh thyme
15g chopped fresh flat-leaf parsley
60ml Marsala
120ml gelée (see page 52) or gelatinous broth (see page 43, step 12) (optional)

In a sauté pan, melt the lard over a medium-high heat. Add the mushrooms in a single, uncrowded layer and season with salt. (If you are going to finish with *gelée*, use a light hand, as the *gelée* can be highly seasoned.) Stir the mushrooms after 1 minute, then add the shallot, increase the heat to high, and cook for 2–3 minutes, until the liquid from the mushrooms has almost completely evaporated. Add the thyme and parsley and turn down the heat to low. Pour in the Marsala and cook for 3 minutes, then add the *gelée*, if using, and cook for another few minutes, until the liquid is almost completely reduced.

Remove from the heat and let the *duxelles* cool to room temperature. Taste for seasoning. Store in an airtight container in the fridge for up to 3 days.

FUNGHI SOTT'OLIO

In this classic Italian preparation, mushrooms are first pickled in vinegar and then preserved in an aromatic oil to produce a mushroom that is at once brightly flavoured yet earthy and unctuous. Stash a jar or two in your pantry when wild mushrooms are copious, then use as an assertive counterpoint to an antipasto of cured or smoked meats, or as a delicious and quick addition to a salad of peppery greens with crispy bacon and goats' cheese. Or eat them, as we often do, straight from the jar. **MAKES ABOUT 1 LITRE**

450g meaty wild mushrooms (such as
    porcini), trimmed and sliced 6mm thick

3 tablespoons fine sea salt

700ml champagne vinegar

500ml extra-virgin olive oil,
    plus more if needed

4 cloves garlic, thinly sliced

1 fresh or dried bay leaf

2 wide strips lemon zest

3 dried cayenne peppers or chiles de árbol

2 or 3 sprigs of oregano

In a bowl, toss the mushrooms with the salt, then transfer to a colander and let drain for 1 hour.

Lightly rinse the mushrooms under cool running water to remove the excess salt. Transfer to a saucepan, add the vinegar, and bring to a simmer over a medium heat. Cook for 5–7 minutes, until tender.

Remove from the heat. Using a slotted spoon, transfer the mushrooms to a clean 1-litre jar. Strain the cooking liquid through a fine-mesh sieve into a second jar or a bottle, let cool, cover, and reserve for use as a stock or in a vinaigrette (see below).

In a saucepan over a low heat, combine the olive oil, garlic, bay, lemon zest, chillies and oregano and cook for 5–7 minutes, until the oil is warm and the aromatics begin to release their aroma. Pour this mixture over the mushrooms. If the aromatics seem crowded at the top of the jar, use a butter knife or chopstick to prod them gently into the depths of the jar. If any mushrooms are poking through the top layer of oil, add additional oil to cover.

Cover the jar and refrigerate for 2–3 days before using to allow the flavours to meld. The mushrooms will stay delicious and vibrant in the fridge for 4–6 weeks.

To serve, let the jar stand at room temperature for 20–30 minutes, then use a clean fork to lift out as many mushroom slices as needed. When you have used up all of the mushrooms, the leftover oil can be saved and used for bruschetta or combined with the reserved vinegar to make a mushroom-flavoured vinaigrette.

2

PROVISIONING THE
LARDER

Imagine the perfect larder in a low, dark corner on the north-facing side of the house. Its clean tiled walls are lined with tidy shelves stocked with jars of suet and drippings; baskets of apples, potatoes, onions, and winter squashes; a tub of golden butter; crocks of sausage and duck confit. Suspended from a hook is a haunch of pork with a thick rind of fat, ready to be turned into creamy lard, a heady broth, perhaps a smoked ham. With these provisions, you could easily whip up an elegant pâté, a hearty pot of soup, or a mess of biscuits. Cool and comforting, the larder affords an assurance that you will be able to provide for your table.

Nowadays a larder might seem like a bit of a throwback. Most contemporary dwellings lack sufficient space for one so if you do have one, consider yourself lucky. Even if you live in a studio flat on the thirtieth floor, it is still worthwhile to sequester a shelf in the fridge for provisioning with meaty supplies.

In the charcuterie, the larder is amply stocked with fats, bone broths, and various types of confits. They are the crucial components of the charcutier's repertoire, the indispensable building blocks for cooking, binding, flavouring, and preservation. They are the meaty staples the charcutier relies on daily – staples you will soon find you cannot do without.

Bones, scraps, and miscellanea that others might toss are hugely valuable in the charcuterie. You get the most from your meat economically, flavourfully, and philosophically when you take a holistic approach to cooking. This is especially true if you are utilising large cuts or whole animals, which will generate a host of meaty by-products that can be turned into essential cooking fats, versatile bone broths, and convenient confits.

Essential Animal Fats

Charcuterie is practically synonymous with fat. Fat provides food with an opulent texture and satisfying flavour. It is necessary for cooking and can play an important role in preservation.

Historically, animal fats were an important part of the human diet and represented a large percentage of our forebear's calorie intake. Then, in the late 1950s, Ancel Keys, an American scientist, promoted the lipid hypothesis, which claimed that there was a direct link between the amount of saturated fat consumed and the incidence of coronary heart disease. Although many dissenters challenged his theories, citing low incidence of heart disease among many traditional populations that consumed large amounts

of animal fats, his ideas took hold. The vegetable oil and processed food industry wasted no time in funding research to support the hypothesis, and animal fats were soon scratched off the menu.

Today we know that the human body's relationship with fat is far more complex than the lipid hypothesis suggested. Good fats will not necessarily make you overweight. Hydrogenated and industrially processed fats are thought to be far worse for the body than high-quality animal fats, and low-fat diets have been associated with increased rates of depression and fatigue. Animal fats contain many nutrients that may actually help the body protect itself from cancer and heart disease. The fat of grass-fed animals has been found to be particularly healthful and to contain high levels of omega-3 fatty acids and conjugated linoleic acids (CLAs), 'good fats' that can improve brain function, reduce inflammation, and promote cardiovascular health.

For proof, look no further than France, birthplace of charcuterie and land of runny cheese, buttery croissants, creamy sauces, and rich pâtés. Overall, the French have a significantly lower incidence of heart disease than Americans. In the Gascony region of France, where duck fat and foie gras is consumed with gusto, it is lower still. So go ahead and spread a little schmaltz on your bread!

Animals store fat in their bodies in a variety of ways. The four basic types of animal fat that can be used in the charcuterie are subcutaneous, intermuscular, intramuscular, and leaf or cavity fat.

Cavity fat is the fat surrounding an animal's organs. It is harvested from the abdominal cavities of whole animals, where it often clings loosely to the organs and body and can be easily peeled or cut away. Cavity fats are excellent when rendered for use in cooking and are generally preferred for baking.

Subcutaneous fat is the fat that resides just below the skin. This sometimes thick layer of fat is plainly evident on the outside of the loins, shoulders, and leg muscles of pork, beef, and lamb. It is this fat, such as the rind of fat on the outside of a pork loin or the fat cap on a rib eye, that helps to keep the muscle meat juicer and more tender during cooking. Some or all of this fat is often trimmed away, however, and in addition to being useful for sausage making, it is also excellent for rendering to use in cooking.

Intermuscular fat, or seam fat, is the layer of fat that separates the muscles from one another. Although less abundant and generally softer than cavity and subcutaneous fat, it is easily trimmed away and can be rendered along with other fats.

Intramuscular fat, or marbling, is the interior fat contained within each muscle. Steaks, roasts, and chops derive their succulence from highly prized intramuscular fat. Since intramuscular fat develops more slowly than other fats, it is most abundant in more mature animals. Although you cannot actually harvest intramuscular fat, except perhaps in the form of pan drippings, it is nonetheless beneficial in cooking and curing meat.

The fat of an animal can convey information about the animal's breeding, diet, and lifestyle. Different breeds of animals can have vastly different fat distribution and meat-to-fat ratios. The Pekin (Long Island) duck is far fattier than its cousin, the lean Muscovy. Traditionally, pigs were sorted into two groups, lard breeds raised primarily for their fat, and bacon breeds selected for meatier purposes. An animal's diet also greatly affects both the quality and quantity of its fat. Carotenoids found in grass can cause the fat of grass-fed beef and lamb to yellow, and although this colour may not be pleasing to the eye, it signals a good diet of healthy pasture. But a similar pigmentation in chicken fat can indicate a primarily corn-based diet, rather than a balanced diet of living grasses, insects, plants, and a variety of grains. The way an animal is raised, the amount

of access it has to the outdoors, and its lifespan all affect fat production. Generally, animals that are adequately exercised, spend part or all of their lives outdoors, and are allowed to grow and mature at a slower pace will have better fat distribution and better-quality fat.

Although we tend to think of butter as a dairy product, it is indeed a preserved fat made from the milk of an animal, mostly the beef cow. Butter has many uses in the charcuterie. It gives duck liver mousse (page 271) a creamy texture, enriches sauces, and can be used to moisten and flavour leaner meats such as chicken. Like other animal fats, butter has had a bad rap in recent years, but we now know that it is far healthier than margarine and a good source for fat-soluble vitamins like vitamins A, D, K, and E. Nutrient-rich, golden butter made from the milk of Jersey cows that have been grazing on green spring pasture is a satisfying revelation.

Here are the different types of animal fats and their uses, followed by instructions on how to render fat properly.

Chicken Fat

A plump, naturally raised chicken will have cavity fat and fat about the neck, tail, and legs that can be easily trimmed away. If you are using a chicken in a preparation in which its skin is not required, the skin is also an excellent source of fat for rendering. Rendered chicken fat is bright yellow, semi-firm at room temperature, and lends a satisfying roast chicken flavour to food. It can be used for sautéing, roasting vegetables, and as a spread for bread.

Duck and Goose Fat

Ducks and geese are migratory birds, genetically structured to stash quantities of fat for their long journeys. Consequently, you can render a good amount of fat from their skin, necks, legs, and cavities. The fat

From left: drippings, strained sausage confit fat, chicken fat, duck fat, foie gras fat, leaf lard

is whiter and firmer than chicken fat and is excellent for sautéing, roasting vegetables, and preparing confits and other preserves.

Pork Fat, Lard, and Leaf Lard
Pork fat from the back, leg, or shoulder, as well as any other subcutaneous or intermuscular fat trimmings, can be rendered to make lard. Properly rendered lard is white, firm, and has a mild pork flavour. This extremely versatile fat can be used for sautéing, roasting, baking, and preparing confits. It is also excellent for deep-frying and has a very high smoke point of 200°C. Foods fried in lard absorb less fat than foods fried in oil, and the lard can be strained and reused several times.

The fat from the pig's abdominal cavity, which is found in two long strips surrounding the kidneys, is rendered to make leaf lard, which is snow white, quite

IN THE CHARCUTERIE

Beef tallow, which can be rendered from nearly any beef fat trimmings, is a useful fat in the kitchen, great for sautéing, roasting, and deep-frying. French fries cooked in bubbling beef tallow are legendary, as is Yorkshire pudding prepared with roast beef drippings or tallow. Like the high-quality cavity fat of the pig, beef suet is found in two long strips in the abdomen, around the kidneys. The rendered fat is used primarily for baking and is the classic choice for the pastry for a mince pie.

Lamb fat, taken from the loin, neck, leg, and shoulder, can also be rendered and used for cooking. It is bright white and firm, like beef tallow, but it has a strong lamb flavour that can be overpowering. It is best used for lamb braises or stews, where it will enhance the already present lamb flavour of the dish.

Butter

Irresistible butter, usually made from fatty cows' milk, is extremely versatile. What isn't made better with a little butter? It can be cut into flour for flaky pastry, used to enrich a sauce, clarified for high-heat cooking, or slathered on good bread and enjoyed in all of its glorious simplicity. From baking and sautéing to roasting and even poaching (butter-poached black truffle, anyone?), butter can do it. In the charcuterie, we often prepare compound butters, which are butters mixed with herbs, garlic, or shallots and sometimes wine. We tuck them under the skin of lean poultry to keep the meat moist during roasting, or dot them on top of grilled or roasted meats in place of a sauce.

Butter has a sweeter and milder flavour than most other animal fats and melts at about body temperature, which accounts for its highly prized melt-in-the-mouth sensation. It is also more perishable than other animal fats and readily absorbs flavours, so is best kept tightly wrapped or in a butter dish and stored in a cool location or refrigerated.

firm, and has a mild pork flavour. It is similar to all other lard in that it can be used for sautéing, roasting, deep-frying, and preserving, but it is most prized for baking, where its brittle, crystalline structure yields flaky pastry and biscuits.

Beef and Lamb Fat, Tallow, and Suet

The terms 'tallow' and 'suet' are used to describe types of both beef and lamb fat. Tallow, the more general term, is used for all rendered fat, and suet refers to cavity fat, both raw and rendered.

Rendering Fat

Rendering is the process of slowly cooking solid raw fats to evaporate their water and extract a versatile, purified fat for cooking. Although you can purchase rendered fats from a good butcher or speciality food store, rendering it yourself is simple and economical, especially if you are already purchasing large cuts or whole animals.

1. **Grind, grate, or chop your fat.** Begin with chilled fat. Grinding fat in a meat grinder will expose more surface area resulting in a higher yield. Ground fat will also render more quickly. (It's a good idea to plan to grind fat for rendering when you are already setting up your grinder for another purpose, such as making sausage, to consolidate tasks.) Cut the fat into 1.5cm cubes and freeze for about 45 minutes. Once the fat has a light freeze, follow the instructions for grinding on page 200, step 5. Alternatively, for small amounts of pork, beef, or lamb fat, grate thoroughly chilled or partially frozen chunks of fat on the large holes of a box grater. For small amounts of poultry fat, chop the fat as finely as possible by hand.

2. **Choose the right pan.** Select an appropriately sized, heavy-bottomed, tall-sided pan. For large amounts (2–4.5kg), use a tall stockpot. For smaller amounts (say, the cavity fat from a duck), a deep saucepan or a Windsor pan (flared sides and flat bottom) will be sufficient. The fat should fill the pan at least halfway but not more than three-quarters. Too much fat in the pan risks splattering and too little or using too shallow a pan risks burning. Add about 2 tablespoons water for every 450g of fat to help hasten the melting process and prevent sticking. The added water will cook off quickly after the fat melts.

3. **Simmer the fat.** Set the pan over a very low heat. Stir the fat every few minutes to prevent it from sticking to the bottom of the pan. If it starts to stick, add a small amount of water to help loosen the fat. Once the fat has melted, turn up the heat to medium and simmer until steam is no longer visible. Cooking times will vary depending on the water content of the fat. Turn off the heat and let the fat cool in the pan for 20–30 minutes. The solids will fall to the bottom of the pot.

4. **Strain the fat.** Ladle the fat through a fine-mesh strainer into a clean glass jar or other container to remove any solids. A mason glass jar is ideal. Keep refrigerated. The chilled fat should be quite firm. However, if it seems loose or watery, return it to the pan and simmer for an additional 20 minutes to evaporate any extra moisture. Properly stored, rendered fat will keep for 3 months in the fridge or up to 6 months in the freezer.

COLLECTING PAN DRIPPINGS

Drippings are the delicious by-products of roasting and searing. They are the fat that literally drips from the meat when you cook a roast, fry bacon, brown a duck breast, or sear foie gras. The fat that is rendered this way will not be as unadulterated as traditionally rendered fat, but it is still worth saving for sautéing greens, roasting root vegetables, or adding flavour to a pot of beans.

To collect the drippings, pour the accumulated fat and juices through a strainer into a clear glass jar or measuring cup. Allow the drippings to cool until the fat separates from the *jus* (juices). Ladle off the fat into another container and save the *jus* for a pan sauce or for enriching soups or braises. Refrigerated, most drippings will keep well for several weeks.

TRUFFLED CREMA DI LARDO

In 2001, we took an extended hiatus from work to eat our way through Spain, Portugal, France, and Italy. But the call of the meat shop was strong, and it wasn't long before we found ourselves working at the famed Antica Macelleria Cecchini in Panzano in Chianti, a picturesque hamlet in the Tuscan countryside. The proprietor, Dario Cecchini, is a truly larger-than-life character for whom food, friendship, music, and the verses of Dante are all equally important. His *crema di lardo*, a gutsy spread made from raw fresh pork fat, rosemary, garlic, and red wine vinegar, was one of the most unique preparations we had encountered on our trip. Raw ground pork fat is worked on a thick marble slab to volumise it, creating a light, delicately textured spread that is eaten on bread in place of olive oil or butter. Making the *crema* at the *macelleria* for the first time with Cher blasting from the stereo, local vino flowing from the carafe, and Dario cheering happily was an unforgettable experience. In tribute, we make this decadent truffle-laden version of the *crema di lardo* each December.

Because you will be consuming it raw, the pork fat must be absolutely fresh. If you can procure chestnut honey vinegar, it adds a certain *je ne sais quoi*. If not, champagne vinegar or any other good-quality wine vinegar will work. Get the most fragrant fresh black truffles money can buy and you won't be sorry. A slick marble surface is optimal and recommended. Other surfaces, such as wood or stainless steel, can discolour or impart unwanted flavours to the fat. Serve the *crema* as part of an antipasto or *salumi* platter with good, crusty bread.

This recipe makes a good bit of *crema*, but since the commencement of the truffle season generally coincides with the holidays, you might consider giving the gift of delicious pork fat laced with truffles to your most deserving and appreciative friends. **MAKES ABOUT 1.2KG**

1.2 kg fresh pork back fat, cut into 2.5cm cubes

1 tablespoon fine sea salt, pulverised in a spice grinder

1 teaspoon freshly ground pepper

1 tablespoon garlic pounded to a paste in a mortar

20ml chestnut honey vinegar or other vinegar

55g fresh black truffle, grated on a rasp-type grater

Chill the back fat in the freezer for 1 hour. Following the instructions for grinding on page 200, step 5, fit the grinder with the smallest plate and grind the fat into a large bowl. Add the salt, pepper, garlic, and 1 tablespoon of the vinegar. Using your hands, mix for about 1 minute. Slowly add the truffle, continuing to mix until it is fully incorporated.

Pour the remaining vinegar on to a large marble slab or countertop. Using your fingertips, spread it over the surface to coat lightly. Turn the fat mixture out on to the marble, then, using your palms, spread the fat over the surface. Massage the mixture roughly with your fingertips, as if washing your hair, then scrape the fat back into a mound and repeat. The idea is not just to mix the fat but also to aerate it. Once the *crema* has tripled in volume, taste for seasoning.

Store refrigerated in a covered jar for up to 5 days. Bring to room temperature before serving.

FLAKY LEAF LARD SAVOURY SCONES

A flaky savoury scone, lightly brown and crunchy on the outside, tender, airy, and moist inside, is the Holy Grail for bakers everywhere. Good savoury scones are the culmination of using quality ingredients in the right proportions, and good techniques. The techniques are fairly simple and easy to master: keep all of your ingredients well chilled, make sure your oven is quite hot, and don't overwork the dough. As for the ingredients, you can't go wrong with all-natural leaf lard that you've rendered yourself. Butter makes a good scone, too, but because butter begins to melt at a lower temperature than lard, any trace of water in the melting butter causes the dough to stick together, rather than separate into thin, distinct flaky layers.

Serve these savoury scones alongside Roasted Nettle Butter Chicken with Spring Vegetables (page 36), or split them and stuff them with browned patties of breakfast sausage (page 208) or griddled slices of Picnic Ham (page 286). **MAKES 14–16 SCONES (6CM IN DIAMETER)**

450g unbleached all-purpose flour, plus
    more for rolling and shaping the dough
2 tablespoons baking powder
1 tablespoon sugar
2 teaspoons fine sea salt
170g chilled leaf lard, cubed
240ml cold buttermilk
1¹/₂ tablespoons unsalted butter, melted

Preheat the oven to 230°C (gas 8). Have ready a large ungreased baking sheet.

In a large, shallow bowl, sift together the flour, baking powder, sugar, and salt. Place the bowl in the fridge to chill for about 20 minutes.

Using a pastry blade, cut the lard into the flour mixture until the mixture resembles coarse crumbs. Make a well in the centre and pour the buttermilk into the well. Using a fork, gradually draw the flour-lard mixture into the buttermilk and mix gently until just incorporated.

Lightly flour a work surface and turn out the dough on to it. Knead gently, just until it holds together and begins to take shape. Pat the dough gingerly into a flat disk. Smooth any large cracks around the edges by pressing inward lightly. Roll out to an even 1.5cm thickness.

Lightly flour a sharp-edged, round 6cm biscuit cutter. Working from the outer edge of the dough towards the centre, cut out as many biscuits as possible, pressing straight down and lifting straight up to make each cut. Lay the scones on the baking sheet, spacing them 0.5cm apart. Gather up any scraps, press them together gently, re-roll the dough once, and cut out more scones (this second round of scones won't rise quite as fully as the first).

Brush the tops of the scones with the butter. Bake for 6–7 minutes, then turn the pan around 180 degrees and continue to bake for about 6 minutes longer, until the scones are lightly browned on top. Remove from the oven and serve immediately.

ROASTED NETTLE BUTTER CHICKEN WITH SPRING VEGETABLES

Compound butters are perfect for infusing lean meats such as chicken, guineafowl, or turkey with a little extra fat and flavour. Tucked under the skin, they baste the bird as it roasts, keeping its meat juicy and flavourful and ensuring golden, crispy skin. We especially like this verdant compound butter made with stinging nettles. If you don't want to gather your supper ingredients, you can often find nettles in the springtime at farmers' markets or speciality produce shops. You want to use the tips and leafy top parts of the plant only. Discard any stems that seem tough or woody. Raw nettles contain a natural acid that will sting your skin and must be lightly cooked before eating. Always wear gloves when gathering or trimming nettles.

This recipe makes more butter than you will need for the chicken. Use the remainder to dress noodles, enrich polenta, or top a steak just before serving. Serve the chicken with a simple green salad. Prepare the chicken the night before you plan to eat it. **SERVES 4, WITH LEFTOVER BUTTER**

NETTLE BUTTER

225g trimmed stinging nettles

1 tablespoon extra-virgin olive oil

2 cloves garlic, pounded to a paste in a mortar

Grated zest and juice of 1 lemon

1^1/$_2$ teaspoons fine sea salt

^{1}/$_2$ teaspoon freshly ground pepper

450g good-quality unsalted butter, at room temperature

CHICKEN

1 roasting chicken, 1.5–2kg

Fine sea salt

10–12 shallots, peeled but left whole

12–14 small new potatoes

12–14 small turnips, trimmed

12–14 young carrots, peeled

2 fennel bulbs, trimmed and each cut into 6 wedges

Bring a large pan of salted water to the boil and carefully blanch the nettles for 2–3 minutes. Drain and cool. The nettles are now safe to handle with bare hands. Squeeze out as much excess water as possible and chop coarsely.

In a food processor, purée the nettles and olive oil until smooth, scraping down the sides of the bowl with a rubber spatula every few seconds. Add the garlic, lemon zest and juice, salt, pepper, and butter and pulse to combine. Taste for seasoning. Reserve 115g of the butter for the chicken and vegetables. Tightly wrap the remainder in cling film and refrigerate for up to 1 week, or freeze for up to 2 months.

Season the chicken inside and out with salt. Fold the wings behind the back (it should appear to be reclining on a deck chair) to prevent the wing tips burning. Using your fingers, gently loosen the skin away from the meat, starting at the top of the breast and working down towards the thighs and drumsticks. Smear about 90g of the reserved butter between the skin and the meat as evenly as possible. Pat the skin back into place. Set the chicken on a tray or platter and refrigerate, uncovered, overnight. (This gives the seasoning time to infuse the bird. Omitting a cover allows the skin to dry out a bit, ensuring better browning.)

Preheat the oven to 200°C (gas 6).

Line the bottom of a roasting tin with the vegetables and dot with the remaining butter. Place the chicken directly on top of the vegetables.

Roast the chicken for 15 minutes, then remove from the oven and baste the vegetables and bird with some of the melted butter that has accumulated in the bottom of the pan. Roast for another 15 minutes, then baste again. Lower the heat to 190°C (gas 5) and continue to roast for about 30 minutes, until a thermometer inserted into

the thickest part of a thigh away from the bone registers 70°C or the juices run clear.

Transfer the chicken to a carving board and let it rest for 10 minutes.

To carve, unfold the wings and, using the tip of a knife, gently remove them from the body. Sever the wing tips from the flats and then separate the flats from the drumettes. Reserve the wing tips for broth. Remove the legs by gently folding the thigh away from the body until the ball joint of the thigh bone pops out of the socket, then use the tip of the knife to cut through the skin and separate the leg fully from the carcass. Cut the thigh away from the drumstick. Carve the meat off the breastbone in a single piece and slice on the diagonal, from the widest part of the breast to where it comes to a point at the bottom. Save the carcass for a broth (see page 41).

According to appetites, put a few slices of breast, half a leg and a piece of wing on each person's plate and adorn with the roasted vegetables.

Versatile Bone Broths

'Indeed, stock is everything in cooking . . . Without it, nothing can be done.'

– AUGUSTE ESCOFFIER

On days when the wind blows a sideways rain against the window, a bowl of beef broth fortifies the soul. If your nose runs and your throat turns sore and scratchy, a rich, ginger-laced duck broth soothes the body. When work goes on too late, or you are too tired to cook, a little roasted pork broth augmented with tiny pasta and grated Parmesan cheese, a poached egg, chopped greens, or a few croutons provides a simple solution for supper. You can dress up a rich game bird broth with a handful of wild mushrooms and a shaving of truffle for an elegant dinner. Or, you can keep it plain and simple and warm yourself up with a mug of chicken broth on a chilly afternoon.

Broths can make all the difference between a good dish and a truly great one. Learning to make excellent-quality broths will add depth and range to your cooking. Broth frequently turns up in the list of recipe ingredients. Braises, soups, sauces, rice, and stews are made better with it. You can use it to moisten roasted vegetables, make a flavourful risotto, or enrich a gratin. In the charcuterie, broth has a number of uses. At the Fatted Calf, we make and use broths daily for poaching, as a gelatinous binder, to enhance the savouriness of our terrines, and to add an intensity of flavour to a variety of meaty preparations. Broth can quickly become a favourite ingredient – one you cannot cook without.

'Soup is a healthy, light, nourishing food, good for all of humanity; it pleases the stomach, stimulates the appetite and prepares the digestion.'

– BRILLAT-SAVARIN

Broths are healthful and healing. It's not just folk wisdom. When you simmer bones, cartilage, and tendons, you extract such minerals as calcium, phosphorus, magnesium, glucosamine, silicon, and sulphur in a form the body can readily assimilate. You also extract gelatine, in evidence in the wobble factor of a cold broth. Gelatine helps your body utilise protein efficiently and promotes good digestion. It is a hydrophilic colloid, a substance that attracts and retains liquids, facilitating digestion by attracting digestive fluids to the stomach. The regular consumption of broth is thought not only to benefit digestion but also to promote strong bones and teeth, prevent joint pain, improve skin condition, reduce inflammation, and help to fight infectious diseases. The nutritional properties and digestibility of bone broth have landed it on the list of homeopathic superfood cure-alls.

Broth makes sense. You can make the most of an animal when you use it to its fullest potential and let nothing go to waste. Bones, hooves, heads, trotters, scraps of meat, even a picked-over carcass or the rib bones from last night's roast are welcome additions to the stockpot. Hooves, feet, and heads contain the most gelatine, and it is good to include at least one of the three when making a broth with good body. At the Fatted Calf, we keep our broths fairly straightforward: meaty bones, heads and feet, water, and a small amount of fine sea salt. This allows us versatility. But at home broths may be altered at whim. Mix pork and duck and throw in a hefty dose of ginger, chillies, and peppercorns for use in curries, noodle soups, or hearty morning rice porridge. A few dried porcini, a sprig of thyme and the green tops of leeks in a broth of chicken or beef makes a hearty *brodo* ready to use in risotto, braises, and stews.

Rich, delicious home-made broths are easy to make. Although they require most of a day to simmer, they are not time-consuming to assemble and require minimal attention during cooking. You can make large quantities and freeze what you will not use right away. If you are feeling ambitious, make a day of it and simmer two or three broths simultaneously. Wispy, meat-scented clouds will steam the windows and your larder will be provisioned with plenty of broth for the season.

Begin with Bones

Chicken backs, wings, heads, and feet, or a whole carcass, along with the heart and gizzard, can be used to produce a mild-flavoured golden broth. It is quite versatile and can be assimilated into many recipes. Interestingly, most conventional chickens, raised in cages, will produce a broth with very little gelatine. Exercise, slow growth and a good diet produce strong bones. Pasture-raised chickens fresh from the farm yield the tastiest, most gelatinous broth. A rooster or 'retired' laying hen is also a good option for broth making.

Duck backs, wings, heads, feet, and hearts, or a whole carcass, will produce richly flavoured deep golden broth. Duck broth is great for risotto, a rich duck *sugo*, and hearty soups such as the Gascon classic *garbure*. In the charcuterie, duck broth can be used to give a flavourful boost to a terrine or as a poaching liquid for a galantine.

Turkey, goose, rabbit, and game birds will all produce a distinctly flavoured, delicious broth. Any bones from the carcass along with heads, feet, and necks can be used to produce an amber brew for use in soups, stews, *sugos*, and risottos. They can also be used to enhance the flavour of terrines and other preparations.

Pork bones produce a rich and sticky amber broth with a pronounced pork flavour. You can use any pork bones, but for the most gelatinous and flavourful broth, add large, meaty shank bones, split trotters, a split head trimmed of its jowls and with its brain

removed, or a little pork skin scraped of fat. Pork broth will add body as well as rich, meaty flavour to braises, soups, and sauces. In the charcuterie, pork broth can be used to braise meats for potted meats, to flavour and bind terrines, and to add body and depth to sauces or similar preparations.

Beef bones are full of gelatine and will yield a rich broth with lots of wobble. Any beef bones are welcome in your stockpot, but a mix of larger meaty shank bones, knuckles, the skinny ends of oxtail, and meaty neck, and rib bones will produce a particularly flavourful broth with good body. Beef broth is the classic choice for French onion soup and for a traditional Provençal daube and is great in other braises, soups, and stews. In the charcuterie, the gelatinous nature of beef broth makes it a useful binding ingredient.

Lamb makes a delicious, dark, rich, and intensely flavoured broth. Meaty shank and neck bones are best, but any lamb bones can be used. The distinct flavour of lamb broth makes it less versatile than other broths, but it is great for use in shepherd's pie, in lamb stews and braises, and in sauces for roasted and grilled lamb. It can also be used to enhance the flavour of lamb terrines.

Preparing Bones for Broth

Break the bones into 8cm pieces. For poultry, use sharp poultry shears to do this. For larger lamb, beef, and pork bones, use a meat saw or, if you are purchasing bones specifically for making broth, ask your butcher to cut them for you. Each 450g of bones will yield roughly 480ml of broth. Rinse the bones with cold water to remove any residue.

Regular or Roasted

Roasting bones before simmering them will produce a darker broth, known as a **brown broth**, with a rich, roasted flavour thanks to the Maillard reaction.

Named after a French chemist, this phenomenon occurs when some foods are heated, causing the molecules in their natural sugars to combine with the amino acids in their proteins, resulting in browning, or caramelisation. You can roast any type of bones from chicken to lamb to produce a broth with a more pronounced rich, meaty flavour.

Whether or not you choose to roast the bones is a matter of personal preference. If you prefer a broth with a deep, roasted flavour, you should roast your bones before simmering them. If you want a more straightforward, versatile broth, skip the roasting to make what is known as a **white broth**. For those who would like a broth with some of the colour and rich roasted flavour but without the intensity of a fully roasted brown broth, roast only half of the bones.

To roast bones for broth, preheat the oven to 220°C (gas 7). Spread the prepared bones in a single layer, without crowding, in a roasting tin or heavy baking sheet. You want plenty of space around each piece to ensure maximum browning. Roast the bones until they are a golden brown. This can take from 45 minutes to over 1 hour, depending on the type of bones and your oven. Remove the tin from the oven and transfer the bones to the pot. If the fond on the bottom of the roasting tin is not overly dark, you can deglaze

it and add it to the pot. Set the warm roasting tin over a low heat and add just enough water to cover the bottom. Using a wooden spoon and some elbow grease, scrape the tin to loosen the fond. Taste the liquid. If it tastes rich and meaty, add it to the stockpot. If you detect any bitterness, discard it.

Adding Aromatics

Bones alone will make a flavourful, meaty, and versatile broth, but there are times when you may want a broth with a more complex flavour. Aromatics such as vegetables, herbs, and spices can be added during cooking to season the broth.

Although you can add a variety of aromatics, keep in mind the stockpot is not the last stop before the compost pile. Do not put anything in the pot that you would not consider eating. Soggy carrots, onion trimmings, wilted herbs, or otherwise forgotten or abused vegetables will yield a poor-quality, tired-tasting broth.

Use aromatics sparingly and add them partway through cooking to avoid a broth that tastes overly vegetal. The flavour of your aromatics should not overwhelm the meatiness of your broth. For example, if you are making a beautiful roasted chicken broth, you do not want celery to be the first thing you taste. Monitor the flavour changes while your broth is simmering. If the aromatics are becoming too prominent, fish them out.

Carrots and other root vegetables can add depth and sweetness to a broth, but they can also cloud it with their starches. If you are aiming for a clear broth, use whole root vegetables and do not simmer them for long periods. An hour should suffice.

A traditional **bouquet garni** consists of bay leaves and thyme and parsley sprigs sandwiched between a half stalk of celery and the fibrous green top of a leek, neatly tied with kitchen twine or packaged in cheesecloth. You can use this same technique for

POST-ROAST BROTHS

Make the most of your roasts with a post-roast broth. The picked-over carcass of a roast duck or chicken, the meaty leg bone of a fresh ham or lamb, or the rib bones from a pork or beef roast all have the potential to be reborn in a broth. On their own, they may not produce the most stellar broth, but they will certainly produce one worthy of a soup. Or, you can combine them with fresh bones for a more flavoursome broth.

whatever herbs and seasonings you want to add to your broth, from fennel fronds and sage sprigs to coriander, ginger, and lemongrass.

Sachet d'épices is the French term for a little bag of spices that steeps in the broth and is fished out when the desired flavour has been achieved. You can use a small muslin bag with a drawstring or cheesecloth tied like a beggar's purse. Any combination of spices can be used, from peppercorns, cloves, and allspice to star anise, cinnamon, and coriander.

Mirepoix is the holy trinity of onions, carrots, and celery that is often added to flavour broths. It has many regional variations and can also include garlic, leeks, celery root, lovage, leaf celery, or various greens. If a darker colour and a caramelised flavour are desired, you can roast the mirepoix before adding it to the broth.

Tomato, which can be added in the form of fresh or roasted tomato, or tomato purée, supplies colour, flavour, and acidity to a broth. The acidity also assists in extracting gelatine by promoting the breakdown of cartilage and connective tissue in the bones.

Onion brûlée, or charred onion, imparts a rich colour and flavour. Generally, an onion is peeled, halved crosswise, and then exposed to intense heat, either cut side down in a sauté pan or cut side up under a grill, until it is blackened. Many Southeast Asian broth recipes call for a similar onion, which is charred whole with the skin intact. After the skin is scorched, it is peeled to reveal a partially caramelised interior. Ginger can be blackened and peeled in a similar manner to add sweet pungency to a broth.

Basic Broth Making

1. **Find the right pot.** Place the bones in a tall, narrow, non-reactive pot. A heavy-bottomed stainless-steel stockpot works best. If you can manage it, have at least two sizes of stockpot in your *batterie de cuisine*: an 11–13-litre pot is great for large batches of broth, and a medium-size 5–6-litre pot is handy for smaller batches. If you make broth frequently, consider investing in a pot with a spigot, or tap, that will allow for easy straining. The bones should fit snugly in your pot, but you will want to leave the top 5–8cm of the pot empty to allow for some expansion and movement during cooking, and for adding aromatics.

2. **Add cold water.** Pour in enough cold water to cover the bones completely. Beginning with cold water helps to produce a clear broth. Certain proteins that help with clarification, such as albumin, will only dissolve in cold water. Leave no bones poking above the surface but use just enough water to cover. Too much water can dilute the flavour of a broth. You might consider using filtered water if your local tap water has flavours you find off-putting. There is no sense in covering your beautiful, delicious bones with water that tastes of chlorine.

3. **Simmer.** Place the stockpot over a medium-high heat and bring to a gentle simmer. Lower the heat to a lazy bubble. An unhurried simmer – a bubble barely breaks the surface every few seconds – helps to produce a clear and tasty broth. A broth that boils turns cloudy and can extract harsh, unwanted flavours from the bones.

4. **Skim.** As particulates rise to the surface, skim them off thoroughly to help ensure a clear and delicious broth. To remove the froth of grey-brown impurities, revolve a small ladle tipped at an angle around the surface of the broth. Skimming several times during the first hour or so is most important. After that, you should not need to skim much, if at all. Do not worry about skimming off

the fat that gathers in shiny pools on the surface. It adds flavour and can be removed easily once the broth has finished cooking and cooling.

5. **Season with salt.** Add just a little salt to your broth. There are definitely two schools of thought on this subject. Many people feel that a broth should not be salted now because you will be reducing it at a later point and you might find yourself with an overly salted reduction. But the benefits of seasoning the broth after skimming it outweigh the unlikely risk of oversalting. Adding a small amount of salt will help to bring out all of the meaty flavours from your bones. (It will also make it easier for you to taste the subtle changes that occur during cooking.)

6. **Add aromatics.** Should you desire the additional flavour of aromatics, they can be added any time after skimming and salting, depending on how much of their flavour and colour you want to impart in the broth. If they are to be an integral part, add them midway. For more subtle flavour, add them for the last hour of simmering. Carefully prod them into place below the surface of the broth, being careful not to disrupt the bones too much.

7. **Do not stir.** If a bone is poking out well above the surface, gently nudge it back into the pot. If the liquid reduces during cooking and the bones begin to poke more than 2.5cm above the surface, add just enough water to cover them again.

8. **Be patient.** Cooking times will vary depending on the type of broth and your particular equipment, altitude, etc. In general, a poultry broth should simmer for 5–6 hours and a pork or beef broth should simmer for 7 hours or more.

9. **Taste for seasoning.** If you taste your broth every hour or so, you will gain an insight into the general progression of flavours. The broth is finished when you determine it has achieved a well-balanced, meaty flavour.

10. **Strain the finished broth.** Especially for large batches, strain first through a colander to remove the bones and other large bits, and then strain a second time through a chinois or other fine-mesh sieve.

11. **Cool.** You want the broth to cool as quickly as possible. The longer it takes to chill, the shorter its shelf life and the greater the chance of spoilage will be. Ladle into clean jars or containers and set in the fridge, uncovered, to chill. As the broth chills, any fat will congeal on its surface. If you will be reducing the broth, or if you do not want the fat left in it, lift off the fat from the chilled broth with a spoon. Cover the broth and refrigerate for up to 5 days, or freeze for up to 3 months.

12. **Reduce.** If you need a concentrated, gelatinous broth for use in a sauce or in the preparation of a terrine, if the broth is a little more watery than you like, or if you want to store broth in a more concentrated form for convenience, you will want to reduce it. It's easy. Just pour the broth into a saucepan and simmer until the desired consistency is achieved. Generally, reducing in volume by one-third or one-half will produce a voluptuous, lip-smacking reduction perfect for saucing a roast, binding a terrine, or adding rich, meaty flavour to any number of dishes.

BASIC RICH BROTH

A basic rich white (unroasted) broth is infinitely versatile, providing a neutral foundation for soups, braises, sauces, and the like. This broth can be made with any one type of bones, or you can try combining different types of poultry, or poultry and pork. Before you begin, review the step-by-step Basic Broth Making directions on page 42. **MAKES ABOUT 6 LITRES**

6kg meaty bones (such as poultry, game bird, rabbit, pork, beef, or lamb), cut into about 8cm pieces
1 teaspoon fine sea salt, or more to taste
Aromatics (optional)

Rinse the bones under cold running water and place in a large stockpot. Add cold water to cover the bones completely. Place the stockpot over a medium-high heat and bring to a gentle simmer. Lower the heat until you achieve a lazy bubble. Using a ladle, skim off the froth as it accumulates on the surface.

After the first hour of simmering, add the 1 teaspoon salt. If you will be adding aromatics, they can be added any time between now and the last hour of cooking. Simmer for 5–6 hours for poultry or about 7 hours for beef, pork, or lamb, tasting every hour or so to monitor the flavour development.

Strain the finished broth first through a colander and then through a chinois or other fine-mesh sieve. Pour into 1–2-litre containers, cool in the fridge and then cover and refrigerate for up to 5 days or freeze for up to 3 months.

BASIC RICH ROASTED BROTH

Roasting the bones before simmering them produces an amber broth with a rich roasted flavour. You can use any single type of bones or a combination of different types of bones you like, but we especially like to use duck bones or a combination of quail, squab, and guineafowl or other game birds to make this earthy, nuanced broth. Before you begin, review the step-by-step Basic Broth Making directions on page 42. **MAKES ABOUT 6 LITRES**

6 kg meaty bones (such as poultry, game bird, rabbit, pork, beef, or lamb, or a combination), cut into about 8cm pieces
1 teaspoon fine sea salt, or more to taste
Aromatics (optional)

Preheat the oven to 220°C (gas 7).

Spread the prepared bones in a single layer, without crowding, in a roasting tin or heavy baking sheet. Roast the bones until they are a golden brown. This can take from 45 minutes to over 1 hour. Remove the pan from the oven and transfer the bones to a large stockpot.

Add cold water to cover the bones completely. Place the stockpot over a medium-high heat and bring to a gentle simmer. Lower the heat until you achieve a lazy bubble. Using a ladle, skim off the froth as it accumulates on the surface.

After the first hour of simmering, add the 1 teaspoon salt. If you will be adding aromatics, they can be added any time between now and the last hour of cooking. Simmer for 5–6 hours for poultry or about 7 hours for beef, pork, or lamb, tasting every hour or so to monitor the flavour development.

Strain the finished broth first through a colander and then through a chinois or other fine-mesh sieve. Pour into 1–2-litre containers, cool down in the fridge, and then cover and refrigerate for up to 5 days or freeze for up to 3 months.

PORK AND DUCK NOODLE SOUP BROTH

When we moved from urban Oakland, California, with its surfeit of amazing noodle joints, to the rural Napa Valley, we found ourselves yearning for those big, steaming bowls of long, tender noodles swimming in a vibrantly seasoned broth. We started making gallons of this all-purpose noodle soup broth to keep on hand for those nights when we are almost too tired to cook. Once you have this broth on hand, all you need to do is toss in some noodles or an egg, a few slices of leftover roast pork or duck, and some chopped green onions, mushrooms, or greens and you have a simple, satisfying meal in almost no time at all. Before you begin, review the step-by-step Basic Broth Making directions on page 42. **MAKES ABOUT 6 LITRES**

3 kg mixed duck bones (such as feet, heads, wings, and backs)

3 kg mixed pork bones (such as hock, trotter, and head), cut into 8cm pieces

Fine sea salt and/or Vietnamese fish sauce

8cm piece fresh ginger, thickly sliced

1 large yellow onion, halved

About 55g Vietnamese yellow rock sugar (available in specialist stores and online or substitute caster sugar)

1 bunch spring onions, green tops only (optional)

3–4 fresh or dried chillies (optional)

Small handful of dried squid or shrimp, lightly rinsed (optional)

Rinse the bones under cold running water and place in a large stockpot. Add cold water to cover the bones completely. Place the stockpot over a medium-high heat and bring to a gentle simmer. Lower the heat until you achieve a lazy bubble. Using a ladle, skim the froth as it accumulates on the surface.

After the first hour of simmering, add a small amount of salt and/or fish sauce to taste. Then, one by one, add the ginger, onion, sugar, onion tops, chillies, and squid, if using, carefully prodding them below the surface of the broth. Simmer for 5–6 hours, tasting every hour or so.

Strain the finished broth first through a colander and then through a chinois or other fine-mesh sieve. Pour into 1–2-litre containers, cool down in the fridge, and then cover and refrigerate for up to 5 days or freeze for up to 3 months.

CRESPELLE AND CHANTERELLE MUSHROOMS IN GAME BIRD BROTH

Tender crêpes stuffed with fragrant wild mushrooms lift a robust game bird broth to a new level of sophistication in this Italian classic. This broth is a perfect *primo* for an elegant meal. Both the crêpes and the filling can be made a day ahead and warmed just before serving. **SERVES 4–6**

4 eggs

180ml whole milk

65g unbleached plain flour

1/2 teaspoon fine sea salt

30g grated Parmesan

FILLING

2 tablespoons unsalted butter

2 tablespoons minced garlic

450g chanterelle mushrooms, trimmed and quartered

Fine sea salt

60ml dry Marsala

1 tablespoon chopped fresh thyme

2 tablespoons chopped fresh flat-leaf parsley

85g grated Parmesan

BROTH

2 tablespoons unsalted butter

3 litres game bird broth (see Basic Rich Broth, page 44)

60ml dry Marsala

Fine sea salt

Chopped fresh flat-leaf parsley or grated Parmesan cheese, to garnish

In a bowl, whisk together the eggs and milk until blended. Slowly whisk in the flour, salt, and Parmesan. Cover and refrigerate for at least 30 minutes.

To make the filling, melt the butter in a sauté pan, over a medium heat. Add the garlic and sauté until lightly coloured. Add the chanterelles and sauté for 6–8 minutes, until they have released their liquid. Season with salt and sauté for 1–2 minutes more, until the chanterelles are tender and lightly browned. Add the Marsala, thyme and parsley and cook for a minute or two, until the pan is nearly dry. Remove from the heat and let cool.

Scrape the contents of the pan on to a cutting board and chop finely. Transfer to a bowl, add the Parmesan, and mix well.

To cook the crêpes, melt the butter in a 15cm frying pan over a low heat. Pour off most of the butter and reserve for later use, leaving behind just enough to coat the pan. Turn up the heat to medium.

Ladle about 60ml of the batter into the pan, tilting and rotating to distribute the batter evenly. Cook for 1–2 minutes, until the top begins to dry and the bottom is golden. Carefully flip the crêpe over and cook the second side for about 30 seconds more. Turn out on to a platter to cool. Brush the pan lightly with some of the reserved butter and continue making crêpes, stacking them as they cook. You should have roughly 12 in total.

Just before serving, bring the broth to a simmer in a saucepan over a medium heat. Add the Marsala and season with salt. Spread the filling on the centre of each crêpe and roll them up. Place 2 or 3 crêpes into each warmed soup bowl. Ladle the hot broth over the top, garnish with parsley or Parmesan and serve immediately.

CREAMY SEMOLINA WITH ROASTED CHICKEN BROTH, GREENS, AND PECORINO

Durum-wheat semolina, a coarse, high-protein flour milled from the endosperm of wheat, is often used for making pasta dough, pizza dough, and breakfast porridge. Whisk a little into roasted chicken broth and you instantly have a silky, savoury soup. Garnished with cooked greens and nutty pecorino cheese, this soup is rustic and satisfying. **SERVES 4–6**

2 tablespoons olive oil

2 tablespoons chopped garlic

1 large bunch Swiss chard or spinach, stemmed and chopped

Fine sea salt

2 litres roasted chicken broth (see Basic Rich Roasted Broth, page 44)

80g durum-wheat semolina

55g unsalted butter, cubed

170g grated pecorino cheese, plus more for sprinkling

In a sauté pan, heat the olive oil over a medium heat. Add the garlic and sauté for about 2 minutes, until fragrant and golden. Add the chard or spinach, season with salt, and cook until wilted and tender. Turn out on to a cutting board and let cool. Lightly press out any excess liquid, then chop finely.

Heat the broth to a lively simmer in a large saucepan over a medium-high heat. While whisking constantly to avoid lumps, slowly sift in the semolina. When all of the semolina has been incorporated, lower the heat and continue to cook, stirring frequently, for about 20 minutes, until the semolina is cooked.

Stir in the butter and cheese, followed by the greens, and continue to stir until thoroughly combined and heated through. Taste and adjust the seasonings if necessary. Ladle into warmed individual bowls, sprinkle with cheese, and serve immediately.

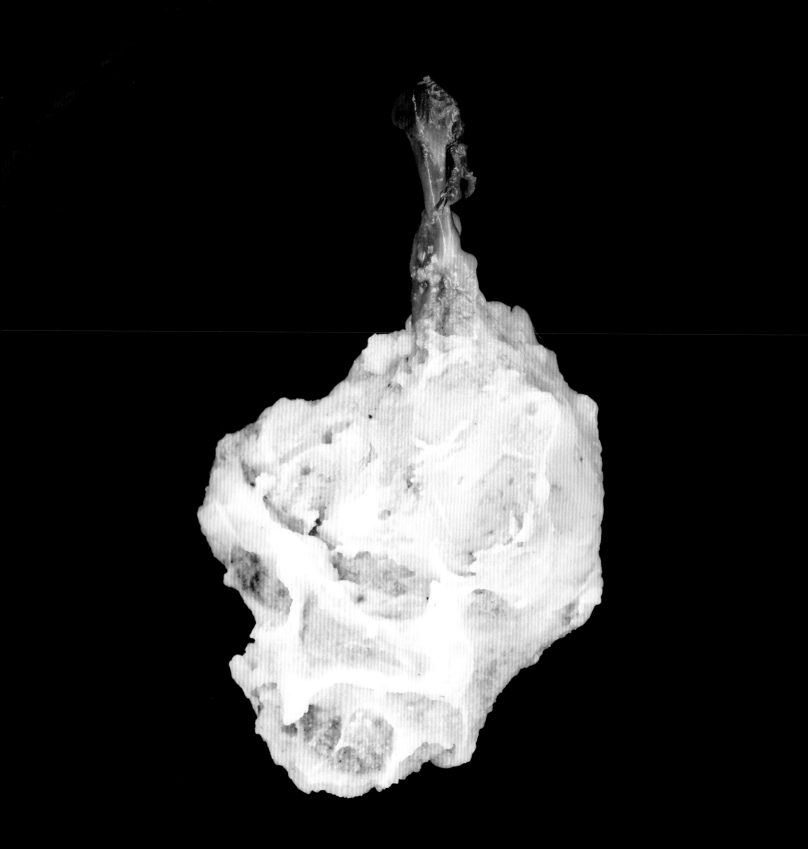

Convenient Confits

No larder is complete without a few jars of confit. It is the ultimate traditional convenience food, providing the foundation for a delicious meal at a moment's notice. Scoop confits straight from the jar and crisp them up in a pan, use them to make rillettes (page 242) and other hors d'oeuvres, or add them to soups, salads, or cassoulet (page 57), *bien sur*!

The process of preparing confits – salting, slow cooking, and sealing meat under a snowy blanket of fat – has been used for centuries with silky, succulent results. *Confire* means 'to preserve'. In addition to being delicious, confits also have the benefit of being long keeping. Before the advent of refrigeration, preparing confits was a way to prevent meat from spoiling. In most cases, the process begins by salting the meat a day ahead of time to add flavour and extract moisture. Then the meat is submerged completely in fat and cooked slowly. Afterwards, the cooked meat is cooled and stored fully shrouded in its cooking fat. The dense fat creates a seal around the meat that prevents air and light from penetrating and keeps bacteria from thriving, holding the meat in a sort of suspended animation. Properly prepared meat confits will keep for many months.

Basic Confit Making

1. **Select the meat.** It is a decadent concept, but almost any meat can be made into confit. Duck, goose, and pork are certainly the most common candidates, but the method can be applied with delicious results to turkey, game birds, rabbit, and even sausages. Cuts that are typically braised, tough cuts, or cuts with lots of connective tissue benefit the most from the long, slow simmering in fat.

2. **Season the meat.** Along with vastly improving the flavour of a confit, seasoning is also key to preservation. Salt draws the moisture out of the meat, lowering its water activity to a level that inhibits the growth of harmful bacteria. In addition to salt, you can also use a variety of spices and herbs to season your confit.

   For most types of confit, you will want to season the meat one day in advance of cooking. (The exception to this would be sausage or another similar highly seasoned meat.) One day ahead, rub the meat with salt and other seasonings and refrigerate overnight. The following day, drain off any accumulated liquid.

3. **Cover with fat.** It is often said that confits are made from meats that are cooked 'in their own fat', but this is either fanciful or a poor understanding of the process. You need a generous amount of fat to make confits, much more than you would likely yield from a single duck or the trimmings from a pork shoulder. When you are making confits, you will want to have plenty of extra fat on hand.

   Generally, recipes for duck confit will call for duck fat, and recipes for pork confit will call for pork fat. But it is not necessary to segregate by animal of origin. For example, you can use pork fat, or even a mix of pork and duck fat, to make turkey or rabbit confit. Just be aware that the fat you use will impart flavour to the meat, so avoid pairing strongly flavoured fats, such as lamb, with mild-flavoured meats, such as duck or pork.

   The fats used to make confits can be reused for several batches, or repurposed for cooking. If you have used a fat for several generations of confit and it begins to appear grainy, separates slightly, and will not become firm when chilled, it is no longer useful for making confits. However, it can still be saved for roasting vegetables or for making rillettes (page 242).

When you are ready to cook your confit, place the meat in a deep, sturdy pan or casserole – preferably one with a fitted lid, large enough to contain the meat in a single layer without too much overlapping. Add the rendered fat, submerging the meat completely. There should be no bones protruding above the surface of the fat.

4. **Cook it slowly.** Most confits will need to cook for several hours at a very low temperature, 110°C–130°C (gas ¼–½). For this reason, it is best to cook a confit in the oven, where it is easier to maintain consistent heat. The meat should cook until it is meltingly tender, that is, until it separates effortlessly with a fork and the meat on a bone pulls away from the bone with little resistance. Once done, remove the pan from the oven and allow the meat to cool in the fat.

5. **Seal it.** The confit can be eaten immediately, but it is better if it is stored under a layer of fat and allowed to ripen. If you plan on using the confit within a week, it can be stored as it is in its cooking vessel, as long as the meat is thoroughly submerged in the fat. For longer storage, carefully strain off the cooking fat into a clear container. As the fat cools, particles and any liquid or *gelée* given off by the meat will settle on the bottom. Transfer the meat to a clean, dry storage vessel, such as a large glass or ceramic jar with a lid. Ladle just the fat (see Harvesting and Using Gelée, right) over the meat until it is completely submerged. Ideally, the fat should top the meat by about 2.5cm. A well-sealed jar of confit may be kept in the fridge for up to 6 months. Label and date your confit containers to help you remember when to use it.

HARVESTING AND USING GELÉE

Gelée is the highly seasoned, rich liquid pulled out of salted meats as they cook slowly in fat. When the fat is strained into a container, the *gelée* will settle in the bottom. Ladle off the fat and use it to seal the confit or for cooking, and pour the *gelée* into a small, clean container. It can be kept refrigerated for up to a week or frozen for up to a few months. In the charcuterie, the *gelée* is highly prized for its powerful, concentrated flavours. At the Fatted Calf, we use the *gelée* harvested from duck confit to season duck liver mousse (page 271) and other terrines. Pork *gelée* makes an excellent binder for roast stuffings and can be use to enhance the flavour of soups, stews, and bean dishes, as well.

SAUSAGE CONFIT

At first glance, preserving sausage in pork fat might seem a bit excessive. The beauty of using this technique is that it actually improves the flavour and texture of the sausage, whereas freezing sausage does no good for either. The relationship between the sausage and the pork fat is symbiotic. The sausage imbues the fat with spices and garlic, resulting in a flavourful fat that can be used to enrich beans, roast potatoes, or sweat vegetables for soup.

You can confit any type of fresh uncooked sausage. They make a great addition to Classic Cassoulet (page 57), and they are wonderful simply browned and served alongside a creamy potato purée. **MAKES ABOUT 4 LITRES**

3 litres lard
1.2kg sausages in pork casings

Preheat the oven to 130°C (gas ½). Melt the lard in a roomy, wide, heavy pan or in a deep, lidded casserole over a low heat. Once it is completely melted, gently drop the sausages into the pot and stir once to prevent them from sticking together. Cover the pan and place it in the oven for 2 hours.

Remove the pan from the oven, take off the lid, and let it cool at room temperature for 30 minutes. Transfer the sausages to large, clean ceramic or glass jars for storage. Strain the fat through a fine-mesh strainer into a clean container and allow it to cool so the *gelée* settles on the bottom. Ladle only the cooking fat over the sausages to cover by about 2.5cm and refrigerate.

Reserve the *gelée* and use as directed on page 52.

Check the fat level after 2–3 hours, as the level can drop as the confit chills. Add more fat if necessary, then cover tightly, label, and date. Submerged in fat, the sausages will keep for up to 6 months.

To use the confit, remove from the fridge and let stand for about 1 hour to allow the fat to soften. Scoop away the top layer of fat and gently pry away the desired number of sausages. If you are not using all the sausages at once, make sure any left in the jar are fully submerged in fat before returning them to the fridge.

WHOLE DUCK CONFIT

Many duck confit recipes call for only the legs, but we prefer to use whole duck. Although you don't actually confit the entire duck – just the wings, legs, and gizzards – using whole birds is a bit more economical and a great way to practise your butchering skills. As a bonus, you also yield bones that can be used to make a delicious duck broth, and fresh duck breasts that are perfect for grilling or searing. This recipe can be easily adapted to suit legs, gizzards, or wings by themselves.

Trust your instincts when making duck confit. Remember that every oven, pan, wing, leg, and gizzard behaves differently. The cooking times included here are rough estimates, so you will need to rely on your senses. Thankfully, duck confit is extremely forgiving. If the wings or legs end up overcooked, they can always be turned into rillettes (page 242), a most delicious error. **MAKES ABOUT 6 LITRES**

2 whole ducks with giblets
1 bunch thyme
3 bay leaves
3 juniper berries
3 litres rendered duck fat, melted

CONFIT SEASONING

2 teaspoons black peppercorns
5 whole allspice
2 cloves
$1/4$ teaspoon finely ground dried ginger
200g fine sea salt

Butcher the ducks following the instructions for Bird on the Bone on pages 70–71. The legs, the first joint of the wings (also known as the drumette), and the gizzards will all be used to prepare the confit. Reserve the remaining parts and use as described above.

Using the tip of a small, sharp knife, remove the silver skin from the gizzards. Weigh the legs, drumettes, and gizzards separately and place them in separate shallow containers large enough to avoid crowding or overlapping. Overcrowding can lead to unevenly seasoned confit.

Prepare the confit seasoning. Finely grind the peppercorns, allspice, and cloves and mix with the ginger and salt. Calculate the amount of confit seasoning you will need for each part. For every 450g of legs and wings, you will need 2 teaspoons confit seasoning. Gizzards, because they are boneless, require only $3/4$ teaspoon per 450g. Sprinkle the confit seasoning evenly over the meat in each container, using a little more on the thicker parts of the legs and drumettes. Once all the seasoning has been dispersed, turn each piece a few times to make sure the seasoning is well distributed. Wrap each container tightly and refrigerate for 12–14 hours.

Preheat the oven to 120°C (gas $1/2$). Quickly rinse each leg and wing under cold running water, just enough to remove any undissolved seasoning. Lightly pat them dry. Drain the gizzards but do not rinse. Place the wings, legs and gizzards in their own deep ovenproof casserole or heavy pan wide enough to accommodate the meat in a single snug layer. Put a few thyme sprigs, 1 bay leaf and 1 juniper berry into each pot and cover the meats completely with the duck fat.

Cover each pot with a lid or tight layer of aluminium foil, and slide all the pans into the oven. After 30 minutes have passed, check each pot to ensure the meat is completely submerged in duck fat. If any pieces are sticking out, add enough fat to submerge.

CONTINUED

WHOLE DUCK CONFIT, continued

After 1½ hours, remove the wings and gizzards from the oven and test to see if they are done. Using a slotted spoon, transfer a wing and a gizzard to a plate and let cool slightly. (Meat straight out of the fat will often feel tough. It is best to give it a moment to cool to ensure an accurate read.) Ideally, the meat on the wings should be soft and pliable but not falling off the bone. The gizzards generally cook at the same rate as the wings. To test, using a paring knife, cut the gizzard in half crosswise. Cut a small sliver from one half and taste. The natural texture of gizzards is quite dense, but it should be tender and not at all rubbery. If either requires longer cooking, return it to the pot, re-cover, and continue cooking for 15–20 minutes longer.

In general, the legs need to cook for about twice as long as the wings and gizzards, anywhere from 2–4 hours depending on their size. The legs are done when the meat starts to pull away from the bone slightly and shreds easily with a fork.

When the wings, gizzards, and legs are done, let them cool to room temperature in their pots. They can now be eaten right away or stored under a layer of fat. Have ready a large, clean ceramic pot or glass jar with a lid. (Alternatively, use several smaller jars.) Ladle fat into the bottom of the pot to a depth of 2.5cm. Using a slotted spoon, remove the legs, gizzards, and wings from their pots and arrange them in the pot. Strain the fat from each pot through a fine-mesh strainer into a single clean container to remove seasonings or other solids. Allow the fat to cool so the *gelée* settles on the bottom. Ladle only the cooking fat over the meats to cover by about 2.5cm, and refrigerate. Reserve the *gelée* and use as directed on page 52.

Check the fat level after 2–3 hours, as the level can drop as the confit chills. Add more fat if necessary, then cover tightly, label, and date. Submerged in fat, the wings, gizzards, and legs will keep for about 6 months, but are generally at their peak of flavour between 1 and 2 months.

To use the confit, remove from the fridge and let stand for about 1 hour to allow the fat to soften. Scoop away the top layer of fat and gently pry away the desired number of pieces. If you are not using all of the pieces at once, make sure that any left in the crock are fully submerged in fat before returning them to the fridge.

CLASSIC CASSOULET

Cassoulet is rustic country cooking at its finest. Its complexity is derived not from overly complicated cooking procedures but from the many layers of flavour contributed by its various ingredients. That said, a well-made cassoulet requires planning and can take the better part of a weekend to assemble. But with one bite of crispy crust, creamy beans, and tender meat, you will know it was time well spent.

The difficult-to-find Tarbais bean is the most commonly used bean for cassoulet in France. Good-quality cannellini beans are a great substitute, however. We have tested the two beans side by side in duelling cassoulets and have found they work equally well. **SERVES 8**

450g dried Tarbais or cannellini beans, picked over and rinsed

225g pork skin with a little fat

$^1/_2$ whole onion, plus 340g minced

1 dried bay leaf

Fine sea salt and freshly ground black pepper

900g crosscut pork hock legs (about 5cm thick)

115g rendered duck fat

70g diced bacon or pancetta, home-made (page 299 or 295) or shop-bought

140g peeled and finely diced carrot

40g finely chopped garlic

500g tomato purée

2 litres duck broth (see Basic Rich Broth, page 44)

1 bouquet garni (bay, thyme, parsley, celery; see page 41)

4 duck confit legs (page 55), separated at the joint into thigh and leg

900g Sausage Confit (page 53; preferably Toulouse sausage)

115g fresh breadcrumbs

To prepare the beans, place them in a large bowl with plenty of cold water and let soak overnight. The next day, drain and discard the water. Place the beans in a saucepan and add the pork skin, the $^1/_2$ onion, bay, and water to cover by at least 5cm. Place over a medium-low heat, bring to a simmer, and cook, uncovered, until almost tender. This can take from $1^1/_2$–3 hours, so check regularly after the first hour to see if they're done.

Season the beans with salt and continue to cook for 20–30 minutes, or until they have a creamy consistency but still retain their shape. Remove from the heat and let the beans cool in their liquor. (The beans can be cooked up to 2 days in advance and stored in their liquor. Cover tightly and refrigerate.)

To prepare the pork hocks, place them in a single layer in a wide, shallow dish and season liberally on both sides with salt and pepper. Cover and refrigerate overnight. The following day, blot the pork dry. In a wide, heavy pan, heat half the duck fat over a medium-high heat. Add as many hocks to the pan as will fit comfortably in a single layer without crowding and brown evenly on all sides. Transfer to a plate and repeat with the remainder of the hocks.

Add the bacon, minced onion, carrot, and garlic to the same pot and sauté over a medium heat for about 5 minutes, stirring occasionally, until the onion is translucent. Season with salt, then add the tomato purée, stir well, and cook for 2 more minutes. Pour in the duck broth and add the bouquet garni and browned shanks along with any resting juices. Bring to a simmer, taste the braising liquor for seasoning, and adjust if necessary. Cover, turn down the heat to low, and simmer gently, stirring occasionally, for $1^1/_2$–2 hours, until the meat is tender and begins to pull away from the bone. Remove from the heat and let cool to room temperature. (The pork can be prepared up to 2 days in advance and stored in its broth. Cover tightly and refrigerate.)

To assemble the cassoulet, preheat the oven to 150°C (gas 2). Drain the beans, reserving 480ml of the cooking liquid and the cooked skin. Dice the

CONTINUED

CLASSIC CASSOULET, continued

skin into 0.5cm squares and set aside. Set the beans, the liquid, and the skin aside separately.

Spoon about half of the beans into the bottom of a *cassole* (earthenware baking dish with flared sides) or other large baking dish. Arrange the duck confit and braised pork on top of the beans and add the reserved braising liquid. Top with the remaining beans. Bake the cassoulet for 1½ hours.

In a large skillet, heat the remaining duck fat over a medium heat. Add the sausages and brown well on all sides. Remove the pan from the heat. Transfer the sausages to a plate. Add the breadcrumbs to the pan and stir to absorb the cooking fat, coating the crumbs evenly.

Remove the cassoulet from the oven and stir very gently. Nestle the browned sausages beneath the surface of the beans. If the top layer seems a little dry, moisten with the reserved bean cooking liquid as needed. Spread the breadcrumbs and pork skin evenly over the top.

Return the cassoulet to the oven and bake for 1–1½ hours longer, until richly browned on the surface. Remove from the oven and let rest for about 15 minutes before serving.

SPECIALS

...hetta $13/lb

...olo Marinated Shoulder $8/lb

...etta Style Shoulder $8/lb

...ntry Rib Roast with
Sweet Green tomato
Chutney $12.50/lb

...HAMS - $12/lb

...BEEF $15/lb

...cken Crepinettes $5/each

...MI Rubbed Pork Shoulder - $8/lb

...ined Pork Shoulder - $8/lb

...rry Pimenton
Pork Shoulder $8.00/lb

...+PORK LINKS $16.00/lb
...AKE'S AMBER ALE

...GANIC

...E BLACK HOG

...MORE PER/lb
...an heritage

SAUSAGES

—All Beef Hot Dogs $12/lb

—Craft Beer Links $12/lb

—Hot Links $12/lb

—Andouille - $12/lb

—Hot Italian - $10/lb

—Breakfast Links - $12/lb

—Sweet Italian - $10/lb

—Mexican Chorizo - $10/lb

—Merguez $12.00/lb

—Toulouse - $10/lb

—Kielbasa - $12/lb

Pâtés & Rillettes

—Pork Rillettes $18/lb oz $9/Jar

—Ciccioli $18/lb oz $9/Jar

—Pâté Maison $19/lb

—Liverwurst $15/lb

—Pâté Rustique $20/lb

—Pork, Bacon, Lobster Mushroom
$22/LB

SALU...

—La Quercia Prosc...

—La Quercia Spec...

—Mortadella $1...

—Salame Cotto $...

—Bierwurst $1...

—Lardo $30.00/lb

—Coppa $3?/lb

—Pancetta $16...

—Guanciale $15/...

—Pancetta Tesa...

3

IN THE
BUTCHER'S

Before you can crank out great sausage, you must first learn how to select quality meats and prepare them properly. To make a gorgeous galantine, you must deftly be able to bone a bird. To master creamy duck liver mousse, you must first clean a heap of duck liver. If you want to create superb charcuterie, you must be able to discern quality meat and possess the ability to perform simple butchery. The butcher's is the place to begin.

Get Thee to a Butcher's

If you value quality (and you should for a variety of reasons detailed in this chapter), skip the meat wing of the supermarket and seek out a good local butcher. A first-rate butcher will source fresh meats from sustainably raised animals, preferably from nearby farms. He or she will have at least some whole animals on hand and will be able to produce the custom cuts you need. If the pork cheek or lamb belly you are looking for is not in stock, a good butcher will help you to procure it. What you get from a good butcher's that you can't get from a chain store is personalised service, attention to sourcing and freshness, and a wealth of information about cutting and cooking meat.

Chefs are fond of saying that a messy workstation equals a messy mind, and we think that dictum holds true for butchers, as well. Although not every butcher is blessed with a flair for fanning out duck breasts or frenching pork racks, an attractive, orderly, well-tended display is often evidence of how much attention a butcher pays the meat on a daily basis. If you want to appraise the quality and freshness of something on display, a good butcher will allow you to examine the meat at close range.

Variety is great, but an oft-rotated small selection of fresh meats will do just fine. Don't expect a good butcher to have everything you need all of the time. In fact, the best butchers will often run out of cuts or won't keep more unusual items in stock on a regular basis. Running out now and then is actually a good thing. It indicates that a strong demand and a fresh supply exist. Learn your butcher's telephone number and remember to call ahead to find out what is great this week or to place special orders for the meats you need.

Shop in advance whenever possible. If you need only a few lamb chops for a weeknight supper, you can usually stop by your local butcher's that same day. But if you are planning a day of sausage making or want to prepare a roast for a dinner party, be sure to plan and shop ahead: you'll need to give your butcher enough notice to gather the items you want, and you'll require plenty of time for both seasoning and prepping.

Quality and expertise cost money. Beware the bargain butcher. In the world of meat, you get what you pay for. If you are budget conscious, a good butcher will be able to offer you choices in your price range – even if they aren't the filet mignon or pork tenderloin you initially had in mind. Untrimmed whole cuts are often more cost-effective than portioned and roast-ready meats. Many of the less glamorous meats, such as shanks and shoulder cuts, require extra time and effort in the kitchen but are extremely delicious when thoughtfully prepared.

Good butchers work for the love of meat and can talk meat almost endlessly. While it pays to do a little homework before entering the shop, a good butcher will be able to steer you towards affordable, delicious cuts that suit your needs. Whether you are curious about the provenance of the meat, need a refresher on cutting and cooking techniques, or just want some new-fangled suggestions for that same old problem of what to fix for dinner, a good butcher will have an answer – or three.

Go Straight to the Source

Good meat comes from good farms. Another great option for supplementing your meat supply is to source meat directly from a local farm that raises animals in a sustainable manner, that is, where animals are pastured, are allowed to move about freely, and are not regularly treated with antibiotics. You can support your community's local farms by joining a meat CSA (community-supported agriculture association), shopping at farmers' markets, or buying a whole animal share. You will often get to interact directly with the farmer and get a pretty good idea of how the animal was raised, what it was fed, and how it was slaughtered.

Keep in mind that you will need to be flexible when you go this route. Most farmers are not butchers, and they cannot offer the huge variety and speciality service that a butcher can. Often, the meat comes in a more 'natural' state: larger cuts, whole animals and birds, or bone-in cuts with skin and fat intact. Buying straight from a farmer is a great excuse to practise your butchery skills in order to get the exact cuts you want.

Meat by Mail

No good butcher's or locally raised meat in your neck of the woods? Well, hallelujah for the World Wide Web. These days you can order everything from Wagyu beef to Red Bourbon turkeys delivered straight to your doorstep, often direct from the farm or slaughterhouse. This is an especially great option if you reside in a remote area and are trying to track down speciality items such as sausage casings, caul fat, or lamb's tongues.

Do-It-Yourself Butchery

For meat enthusiasts, butchering is fun, educational, and economical. With a little planning and effort, you can yield all of the cuts you want for charcuterie and then some. Whatever type of home butchery you choose to undertake, here are a few things to keep in mind.

Start small. This is especially true if you are just beginning to wield a butcher's knife. Small creatures, such as poultry and rabbit, have relatively simple skeletal structures and can be easily butchered, with minimal effort. Once you have a little experience under your belt, you might advance to a pork primal, a goat, or even half a hog.

Sharpen your knives. Using dull knives makes for tedious and potentially dangerous work. For large butchery projects, make sure that your saw blades are sharp and replace if necessary.

Make a plan. If you aspire to cut up a whole hog to produce your own bacon, sausages, and terrines, a plan of action is crucial. You don't want to end up with more roasting cuts than you can reasonably use or store. Likewise, you don't want to find yourself with more sausage meat and curing cuts than you can realistically process in a timely fashion. Write yourself a 'cut sheet', a list of all of the cuts you want to butcher along with the corresponding amounts in order to help you to stay organised. If you plan on brining or dry curing any of the cuts, such as picnic hams or bacon, make the brines and rubs in advance.

Prep your home butchery. For small projects such as duck or rabbit, procure a dedicated meat-cutting board that is at least twice as large as the carcass to allow you to cut comfortably. For larger projects use a freestanding butcher's block or sturdy worktable topped with two or more large cutting boards. Lay a tarp under your work area for easier cleaning up. (Tarps are also handy for transporting the whole beast.) Regardless of the size of your butchering project, set up all of the necessary equipment beforehand. Have on hand cling film, butcher's paper, tape, large freezer bags if you will be freezing any cuts, aprons, towels, butcher's twine, knives, and a saw, if necessary. Refrigeration or freezer space must be readily available, or at least a few ice chests and plenty of ice. It's good to have two or more containers on hand for cuts and scraps. If you will be generating bones, have your stockpot or roasting pan nearby.

Work together. Cutting up a whole pig, lamb, or goat is a big project that will yield a lot of meat. To lighten the workload, consider joining forces with other meat enthusiasts with whom you can share the fruits of your labours.

KNIVES YOU REALLY NEED

Whether you're breaking down a whole hog to make bacon and salami or just breaking down a chicken to braise for supper, a few sharp, well-made knives will help you to get the job done safely and efficiently.

Butcher's saw: Available in a variety of lengths (usually 40–64cm), a sturdy butcher's saw is designed to cut through large bones, making it indispensable for whole-carcass butchery.

Boning knife: This is probably the most utilised knife in the butcher's. Its narrow, medium-length blade, which generally curves inward, is used to remove meat and poultry from the bone, and its sharp, tapered point is used to trim fat and cut tendons.

Butcher's knife or scimitar: The long, substantial blade with a curved silhouette is used to portion meats and slice through skin, cartilage, and small bones.

Chef's knife: A versatile workhorse with a long, wide blade, the chef's knife is used for slicing and chopping of all kinds.

Meat cleaver/chopper: The cleaver's heavy, wide rectangular blade is designed to cut through bones and larger pieces of meat.

Poultry shears: The curved blades of these shears allow you to cut safely and easily through the joints and bones of poultry without tearing the skin and flesh.

Rubber mallet: This primitive but extraordinarily useful tool is good for tapping a cleaver safely through hard-to-break bones and for flattening pork bellies for pancetta (page 295).

Sharpening steel: This long-handled rod of steel or ceramic helps maintain an edge between sharpening blades with a sharpening stone.

Sharpening stone: This simple, traditional tool, also known as a whetstone, is used to restore the blade of any knife when its edge becomes dull.

From left: poultry shears, sharpening steel,
meat cleaver, sausage knife, butcher's knife,
chef's knife, utility knife, paring knife

Birds

Who can resist the aroma of succulent roast duck, its crispy skin glistening? Who can say no to a chicken simmering in a pot, effusing its comforting aroma? From plump little quail sizzling on the grill to a grand, gratitude-inspiring turkey on the Christmas table, the world of poultry is diverse and delicious.

Birds were first domesticated some ten thousand years ago. Many archaeologists believe the first domesticated bird, the chicken, may have been reared for sport, not supper. But it didn't take long for humans to catch on to the delicious dishes they were missing, and soon enough chickens, ducks, and other poultry were being raised for both eggs and meat. Birds were popular with the ancient Romans, whose culinary innovations included the omelette as well as stuffing birds for cooking.

Until the 1950s, most poultry was raised on small family farms where the birds scratched, pecked, and hunted for a fair ration of their own food. Then the invention of antibiotic- and vitamin-fortified feed revolutionised poultry farming, allowing farmers to move their flocks into vast warehouses that could turn out large numbers of birds in short periods of time. Through rigorous breeding and intensive feeding and confinement methods, factory farms were soon able to transform a day-old chick into a 2.5kg chicken in less than a month and a half. Poultry became a product: cheap, fast, and not terribly delicious.

But a backlash against these bionic birds is under way. Many small farms are returning to raising poultry the old-fashioned way and putting the birds back outside. On the home front, more and more urbanites and suburbanites are donning their straw hats and trying their hand at raising a flock, too. Although there is a cost associated with pastured poultry – raised outdoors and fed a natural diet – the results are well worth it.

Birds in the Butcher's

A good butcher will offer at least a small selection of birds. Poultry is fairly perishable and is ideally purchased and used within a week of processing (the technical term for killing and preparing an animal for sale). Source quality poultry from butchers' that have a high turnover or that will special-order meats they do not regularly stock. Tracking down quail, guineafowl, squab pigeons, geese, and even game hens may take a little research. Asian meat markets are often a good source for quail, squab, and duck if your regular butcher cannot procure them.

When purchasing poultry, check that the bird is fully intact, with no broken wings or bones, and is free of bruises or blemishes, which can indicate poor handling during processing. Buy birds with their skin on. The skin does not have to be monochromatic, but it should look lively and fresh. Don't be turned off by dark skin, especially on squab pigeons or heritage breed turkeys, which may have purple or brown coloration on the neck, wings, or legs. This is a natural occurrence.

Quail are little birds with loads of flavour. Most domestic quail are Coturnix quail (Japanese quail) that can be further subdivided into breeds such as Golden Range, Tuxedo, Scarlett, Manchurian Golden, or Rosetta. Ready-to-cook quail are often sorted into small, medium, and large and are usually available whole (on the bone) or semi-boneless (glove-boned), meaning that all of the bones, except the tiny wing and leg bones, have been removed. Quail are generally priced by the piece rather than by weight. An adult Coturnix quail weighs 100–140g.

These small birds are nearly always cooked whole. If you are serving quail as a first course or appetiser, or if you are stuffing quail for a main course (as in Fig-and-Sausage Stuffed Quail, page 185), allow one quail per person. For a main course of unstuffed quail, allow two quail per person. Quail take brines and marinades quite well and are excellent stuffed with sausage, cornbread, wild rice, and grains. When stuffing quail, a quick three-hour brine is recommended to keep them from drying out, since the quail meat cooks faster than the stuffing. The versatile quail can be grilled, pan seared, roasted, or fried, generally at a high temperature to achieve crispy skin with a juicy interior.

Poussins are actually just immature chickens. They are usually sold whole by weight, and generally weigh between 400 and 450g, but no more than 750g. You will need two poussins for two servings.

These small but stocky birds are usually cooked whole and can be roasted, grilled, or fried the same way you would any chicken, though for a shorter cooking time because of their smaller size. One popular cooking method, known as *al mattone* (with a brick), is to spatchcock the bird (remove the backbone and flatten the bird), then roast or grill it under a tin foil-wrapped brick or other heavy weight. Poussins adapt well to almost any type of seasoning and marination, including citrus, garlic, most herbs and spices, and chillies. They can be substituted for chicken when making sausages and pâtés.

Squab pigeon, which is a domesticated fledgling pigeon, is a little bird with richly flavoured meat and a deep rosy colour. It is one of the oldest domesticated birds, popularised in ancient Rome and Egypt, feasted on by nobility throughout medieval Europe, and perfectly presented on modern-day plates in the world's finest dining establishments. Don't let this tasty little bird's haughty history deter you from enjoying its powerful, succulent flavour. The birds typically weigh 450–500g and are generally sold whole, with head and feet attached.

Although squabs can be cooked whole, they are frequently divided into breasts, the most prized meat and generally preferred rare, and legs, which are best cooked a little longer. The breasts can be seared, roasted, or grilled, and the legs are good turned into a confit, braised, or boned and used for sausage along with a little pork fat. Be sure to save any bones for a rich game broth (see page 44).

Chicken is the most common poultry in the world, a descendant of the red junglefowl, domesticated thousands of years ago in South Asia. Chicken has become so ubiquitous in recent years that it has taken a lowly place at the culinary table. True, a factory-farmed bird is an insipid and overused foodstuff, but a pasture-raised bird that has spent its life frolicking and nibbling on natural grass and insects yields juicy, irresistible meat with a slightly nutty flavour.

In the charcuterie, we always purchase whole fresh birds directly from a local farm to ensure the highest quality. Whole birds yield not only meat but also a heap of bones, heads, and feet for the stockpot and, if you are lucky, tasty giblets including jewel-like livers that can be grilled or used in Chicken Liver Crostini (page 80).

Guineafowl is a remarkable-looking, rambunctious bird native to the African savannah. Many farmers incorporate guineafowl into their pastured chicken flock to act as a protector against predators. The meat is earthy and darker than chicken meat but not quite as rich as duck.

Young guineafowl is delicious roasted whole, while mature birds make excellent braises or stews. The breasts and the legs can also be cooked separately. In the charcuterie, guineafowl can be used to produce sausage and pâtés and is a fitting substitute for more unusual game birds, such as partridge and grouse.

Duck is the pig of the poultry world, extremely versatile in its uses, especially in the charcuterie. Domestic breeds of duck, most of which originated in Asia, have been farmed for thousands of years for their rich eggs, savoury meat, and sometimes for their fattened livers (foie gras).

Most ducks weigh 1.8–2.3kg. A whole roasted duck with its crispy, mahogany-coloured skin is a sight to behold. But, like many other fowl, the breast and leg meat cook at different rates and better results can often be achieved by using different cooking methods. The large, tender breast, the flavour of which is often compared to steak, is best cooked rare to medium-rare on the grill or in a pan. The legs can be slow-cooked, either as a braise or confit. The skin of the duck can be used to line terrines. It is also outfitted with a generous layer of subcutaneous fat that can be rendered for cooking (see page 30). To release some of the fat during roasting or searing, prick or score the skin. If you are in luck, your duck will come with its gizzard (great for confit), heart (good for grilling), neck (save for *Cou Farci*, page 266), and liver (perfect for duck liver mousse, page 271). Duck bones, heads, and feet make a hearty broth (see page 45), perfect for soups and braises.

Some ducks are also specially raised to produce foie gras, excellent for *torchons*, terrines, mousses, or simply searing. In addition, ducks harvested for foie gras produce both extra-large legs, which are great for confit, and breasts, referred to as *magrets*, which can be cooked, smoked, or cured.

Goose meat is darker and more intensely flavoured than turkey, and fattier and gamier than duck. The Embden, Toulouse, and Chinese are the most common breeds available. A suitable goose for roasting will generally weigh somewhere between 4.5 and 6.3kg, while a mature goose, perfect for charcuterie, will weigh between 8 and 8.5kg. Roast goose is a traditional Christmas treat, but goose also makes fabulous confit (follow the directions for Whole Duck Confit, page 55). With its thick layer of subcutaneous fat, goose has plenty extra for rendering (see page 30), and the carcass can also be used to make a rich broth (see page 44). Geese are also raised for foie

DARK MEAT OR WHITE?

Dark meat is what avian biologists refer to as 'red muscle'. It is the muscle used for sustained activity, such as walking in the case of a chicken. White meat or 'white muscle' is muscle that is used only for short bursts of activity such as, for chickens, flying. Birds such as chickens and turkeys that are better suited to walking than flying will have dark leg and thigh meat and white breast meat, while birds such as ducks and geese that have breast muscle capable of sustained flight will have red muscle or dark meat, throughout.

gras. In fact, the *foie d'oie* is even fattier than duck foie gras.

Turkey, a cousin of the pheasant, is an American original. Today's domesticated turkeys are the descendants of wild turkeys indigenous to North and Central America. This impressive giant of the poultry world ranges in size from small, weighing in at 2.3–4.5kg, upward to 18kg. The most common size is the young roaster, which weighs 4.5–9kg and is 5–7 months old.

Because turkey is extremely lean, brining it is highly recommended. Roasting is the most common cooking method, but a good case could be made for braising the legs or turning them into a confit. Separating the legs from the breast prior to roasting is another good option, as it allows you to cook each perfectly. A turkey often comes with its giblets and neck, which make delicious additions to turkey gravy or turkey broth, along with the raw or roasted carcass. For the most flavourful turkeys, look for heritage breeds such as the Norfolk Black, Bronze, or Bourbon Red, raised on small farms with access to pasture.

Basic Bird Butchery

Birds are a great place to begin to learn butchery, and knowing how to butcher a bird is a skill that will come in handy again and again. All birds have a similar skeletal structure and can be butchered using the same methods. Once you have mastered breaking down a duck or chicken, you can use those same techniques for other poultry.

Bird on the Bone

When braising, frying, or grilling a bird, you usually want to cut the bird into four to eight pieces, leaving the meat on the bone as it will be more flavoursome and juicy. Be sure to save the backbone for the stockpot. See instructions below.

Boned Bird

In the charcuterie, when you are preparing sausage, pâtés, or the like, you generally want a bird that is completely boneless. Many recipes, such as Duck and Lemongrass Sausage Patties (page 204), call for just the meat of the bird. The method illustrated on pages 72–3 allows you to separate the meat fully from the whole carcass, yielding meat, bones, and wings.

Whole Boned Bird

Some recipes, such as Duck Stuffed with Farro, Figs, and Hazelnuts (page 190), call for a whole boned bird – that is to say, a bird that has had its bones completely removed, but the meat left intact. Others, such as the Veal and Chicken Galantine (page 265–6), call for the intact skin of a whole, deboned bird. The method illustrated on pages 74–5 should be used in those instances.

BIRD ON THE BONE

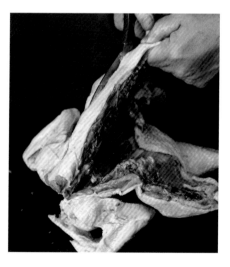

◄ 1. Remove the backbone. Place the bird, breast side down, on a cutting board. Using poultry shears or a sharp, heavy knife, cut along either side of the backbone to remove it.

▶ 2. Cut through the breastbone. Cut the bird in half lengthwise, severing it through the breastbone. You should have two equal halves..

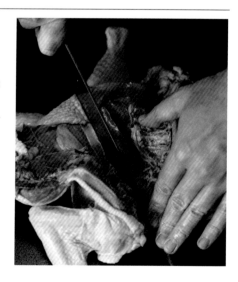

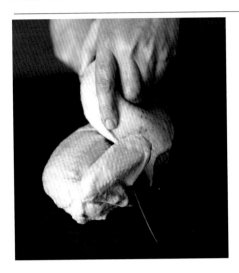

◀ **3. Separate the legs from the breasts.** Place a half, skin side up, on the board. About 2.5cm below where the breast comes to a point, sever the breast from the leg by cutting through the meat and bone. Repeat with the other half of the bird.

▶ **4. Cut the thighs away from the drumsticks.** If you wish to cut the legs into two pieces, cut through the joint between the thigh and the drumstick.

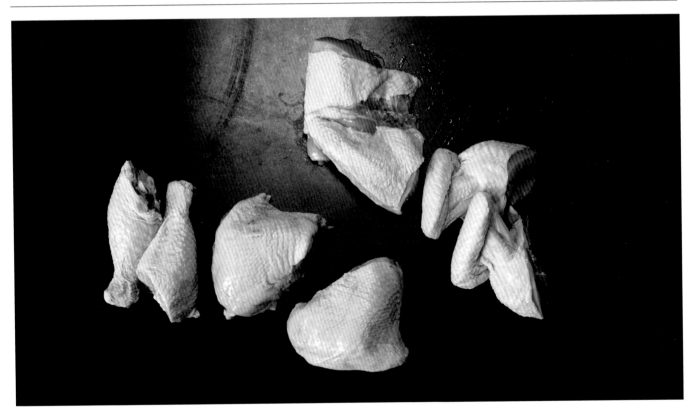

5. Halve the breasts (optional). If you want a bird cut into eight pieces (as pictured above), cut each breast in two pieces. Place a breast half, skin side up, on the board. Starting about 2.5cm below the wing joint, cut down through the breastbone.

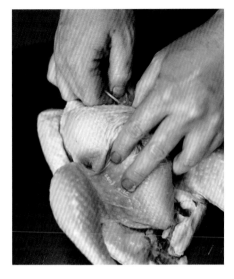

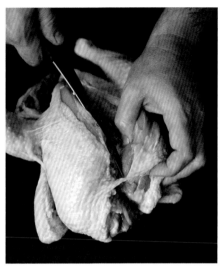

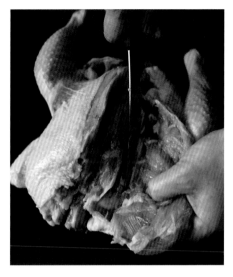

1. Remove the wishbone. Place the bird, breast side up, on a cutting board. Using your fingertips, locate the wishbone underneath the meat at the top of the breasts (neck cavity end) and wiggle it free.

2. Remove the breast from the breastbone. Feel where the breastbone, or keel, runs down the middle of the breast. Using a boning knife, make a cut directly on one side of the breastbone, extending it from the top of the breast all the way to the bottom. You want the tip of the knife to be pressing directly on the bone.

3. Carve the meat off the breastbone, working away from the bone and peeling the meat back with the fingers of your other hand as you go. Continue until you have removed the entire breast. Repeat on the other side of the breastbone.

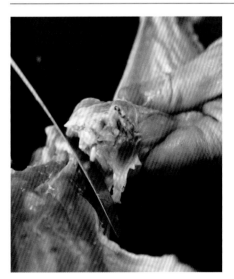

◀ **4. Remove the wings.** Grasp a wing and extend it away from the body. Then, using a boning knife, cut through the shoulder joint, severing the wing. Repeat on the opposite side. Reserve the wings for another use.

▶ **5. Remove the legs.** Hold the carcass down with one hand while gripping a leg with the other. Push the leg backward until you feel the ball joint at the top of the leg pop out of its socket. Using the tip of your boning knife, scoop out the 'oyster', the small muscle that sits next to the backbone and above where the leg popped out of the socket. Continue to cut around the leg bone, following the back all the way down to the tail. Repeat with the other leg.

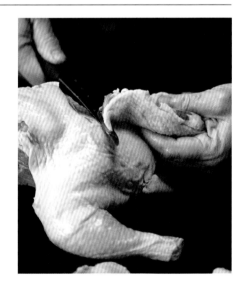

IN THE CHARCUTERIE

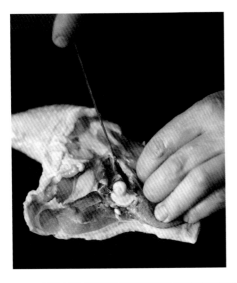

◀ **6. Cut the leg meat away from the bone.** Lay the legs, skin side down, on the board. Using the boning knife, slice along each side of the thighbone, pressing against the bone as you go.

▶ **7. Slip the knife underneath the thighbone** to free it from the muscle, scraping any extra meat away from it as you go. Then do the same with the drumstick.

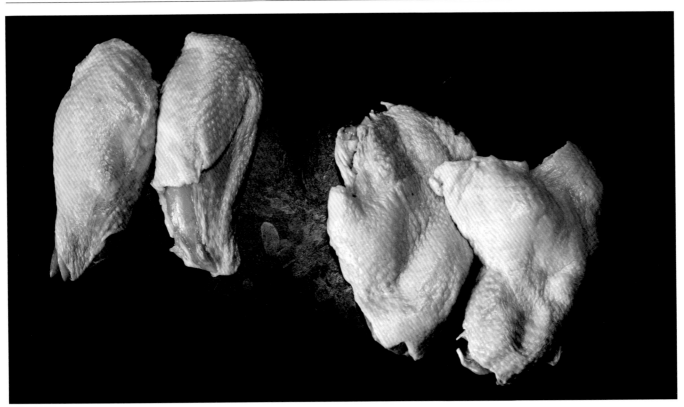

8. The result: Boneless breasts (left) and legs (right).

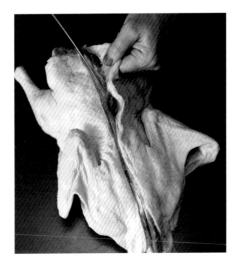

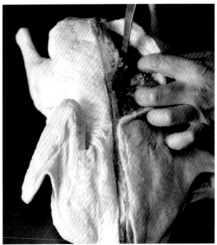

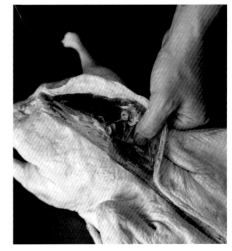

1. Cut through to the backbone. Place the bird breast side down on your cutting board. Using a boning knife, make a cut down to the backbone from the neck to the tail end.

2. Separate the thighs from the backbone. Choose one side of the bird to bone first. Using your fingers, locate the socket where the thigh bone attaches to the backbone. Insert the tip of your boning knife into the socket to separate the thigh from the back.

3. Separate the wings from the backbone. Repeat this process with the wing joint, cutting at the point where the drumette connects to the wing joint.

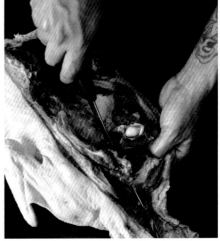

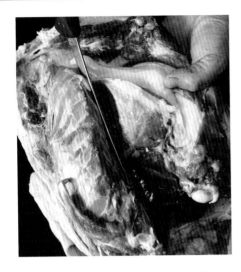

4. Continue to carve the breast and thigh meat away from the carcass.
Using a boning knife, carve the breast and leg meat away from the rib cage and backbone.

5. Debone the other side. Repeat steps 2 to 4 on the other side of the bird.

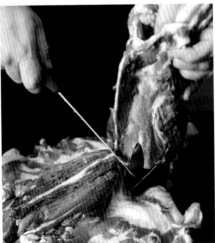

6. Detach the carcass. At this point, the meat should only be attached to the carcass at the breastbone. Leaving the meat flat on the board, peel the carcass away with one hand while cutting against the breastbone with the other to free it completely.

7. Sever the wing tips and flats away from the drumettes by cutting through the bones at the joint.

CONTINUED

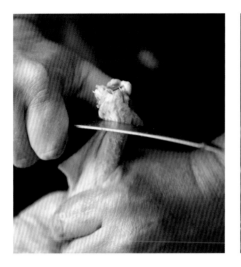

8. Debone the wings. Make a circular cut around the tips of the drumettes to completely sever the tendons. This will make it easier to remove the bones from the bird without making a large hole.

9. Expose and loosen the drumette bones by cutting directly against the bones.

10. Holding the wings tightly, use your thumbs to push the drumette bones out of the meat, leaving a small hole behind on either side of the bird. If there is resistance or the meat starts to tear, cut the meat from the bone where it is still attached.

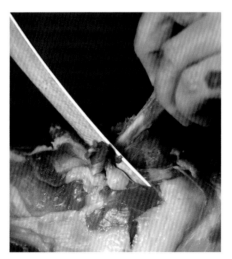

◀ **11. Remove the thighbones.** Carve along both sides of the thighbones with your boning knife, then cut along the undersides to loosen the bones. Sever the thighbones from the drumstick at the joint, then carve the bone away completely.

▶ **12. Debone the drumsticks.** Remove the bone from the drumsticks in the same manner as the drumette, by scoring the outside of the leg to sever all the tendons. Neatly carve away as much meat as possible from the bone on the inside of the leg, then push the bones out through the hole.

IN THE CHARCUTERIE

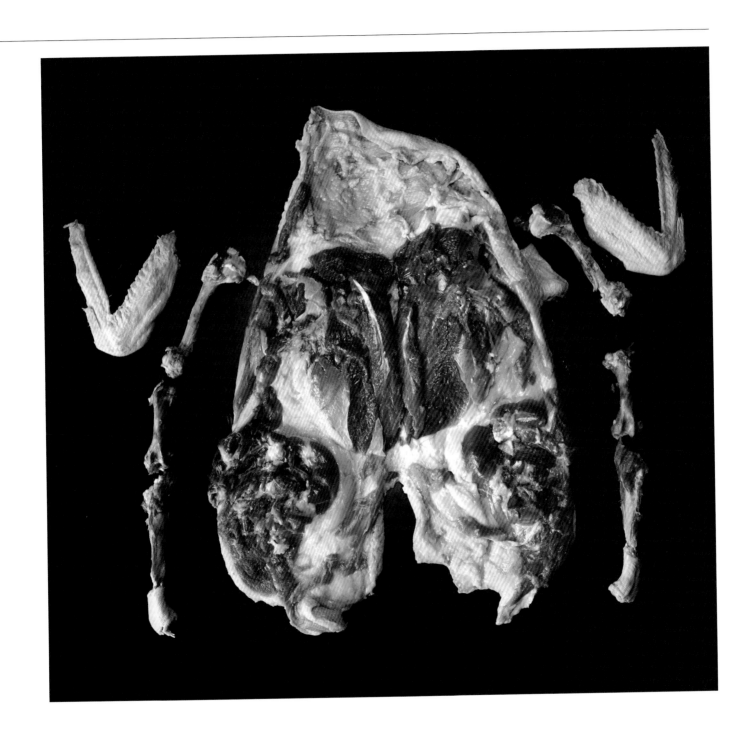

GINGERY BRAISED DUCK LEGS

Although duck legs can be a bit chewy when roasted, braising renders them tender and succulent. This comforting stew has a bold dose of ginger, a perfect foil for the richness of the duck. We generally eat this braise over egg noodles, but it can also be served over fragrant jasmine rice. Should any duck go uneaten, the meat can be shredded the next day and folded into fried rice. It is better to use a roasting tin with a rack to keep the bird off the base of the pan and allow air to circulate. **SERVES 6**

6 duck legs
1 tablespoon fine sea salt
1 tablespoon Five Spice (page 11)
2 teaspoons sugar
1/2 teaspoon freshly ground pepper
1 teaspoon toasted sesame oil
2 tablespoons soy sauce
2 tablespoons Shaoxing (rice) wine
    or dry sherry
3 tablespoons grated fresh ginger
1 tablespoon finely chopped garlic
1 tablespoon rendered duck fat
2 litres Pork and Duck Noodle Soup Broth
    (page 45) or duck broth (see Basic Rich
    Broth, page 44)

TO SERVE
450g fresh Chinese egg noodles, cooked
    according to the packet instructions
100g thinly sliced spring onions
150g beansprouts
15g fresh coriander leaves

Lay the duck legs in a baking dish large enough to accommodate them in a single layer. Season them on both sides with the salt. In a small bowl, stir together the Five Spice, sugar, pepper, sesame oil, soy sauce, wine, ginger, and garlic. Pour the mixture over the duck legs and roll them in the marinade to coat evenly. Cover and refrigerate for at least 4 hours or overnight.

Remove the duck from the fridge. Heat a large, deep frying or sauté pan over a medium heat and melt the duck fat, tilting the pan to coat it evenly. Remove the duck legs from the marinade, reserving the marinade, and pat the legs dry. Place the legs, skin side down, in the hot pan and cook slowly to brown the skin and render some of its fat. When the skin is a deep gold, turn the legs and continue to brown on the meat side for 2–3 minutes.

Tilt the pan and ladle off as much of the cooking fat as possible. Save this fat for later use. Add the broth to the pan along with the reserved marinade. Bring to a slow simmer, then lower the heat slightly and cover. Cook for 2–2¹/₂ hours, until the meat is tender and begins to loosen from the bone.

Remove the pan from the heat. To crisp the duck legs, preheat the oven to 230°C (gas 8). Transfer the legs from the braising liquor to the rack of a roasting tin, arranging them skin side up. Blot the legs dry with a paper towel, then brush them liberally with the saved cooking fat. (Reserve any remaining fat for sautéing or stir-frying vegetables.) Place the legs in the oven and roast for about 10 minutes, until brown and crisp.

To serve, heat the braising liquor. Divide the cooked noodles evenly among large, deep soup bowls and add enough braising liquor to each bowl to just cover the noodles. Rest a crispy duck leg on the noodles in each bowl and garnish with spring onions, beansprouts, and coriander leaves.

CHOPPED CHICKEN LIVER CROSTINI

If you are lucky enough to have the liver included with your whole chicken, it is like the prize in the bottom of the box. You can use chicken livers to garnish terrines, enrich a sauce, or as hors d'oeuvres. The texture of fresh livers is far superior to that of frozen, so do not be tempted to buy the latter. Ask your butcher to set aside any extra livers when fresh chickens arrive. If chicken livers are unavailable, duck livers make a delicious alternative for these crostini. Serve as part of an antipasto assortment or alongside a green salad. **SERVES 6**

450g chicken or duck livers, cleaned and trimmed as directed on page 269
Fine sea salt and freshly ground black pepper
140g diced pancetta or bacon, home-made (page 295 or 299) or store-bought
Olive oil, for cooking, binding, and brushing
40g minced shallot
1 clove garlic, minced
1/2 teaspoon chilli flakes
2 tablespoons dry Marsala
2 tablespoons finely chopped fresh marjoram or oregano
1 tablespoon finely chopped fresh thyme
40g salt-packed capers, rinsed, soaked in water to cover for 20 minutes, drained, and chopped
1 baguette

Season the livers with salt and pepper. Cover and refrigerate for 2–4 hours.

Heat a large sauté pan over a medium heat. Add the pancetta and allow it to brown slowly, releasing its fat. When most of the fat has rendered, using a slotted spoon, transfer the pancetta to a plate and reserve.

There should be enough fat in the pan to coat the bottom evenly with a thin layer. If your pan is shy of fat, supplement with a splash of olive oil. Increase the heat slightly and add the livers in a single, uncrowded layer. Brown the livers, turning once, for about 1½ minutes on each side. They should be cooked through but still a touch pink in the middle. Transfer the livers to a cutting board to cool.

Add about 2 tablespoons olive oil to the pan and lower the heat slightly. Add the shallot and garlic to the pan and sauté until golden. Add the chilli flakes, then deglaze the pan with the Marsala, dislodging any browned bits on the pan bottom with a wooden spoon. Scrape the contents of the pan into a bowl.

When the livers have cooled slightly, chop them coarsely with a chef's knife, making several passes over them until they achieve a somewhat uniform consistency. Add the livers to the shallot-garlic mixture along with the marjoram, thyme, capers, and reserved pancetta. Stir in 60ml olive oil, or more, to taste.

Preheat the oven to 190°C (gas 5). Cut the baguette on the diagonal into slices 12mm thick. Arrange the slices in a single layer on a baking sheet. Brush the tops with olive oil.

Toast the slices for about 8–10 minutes, until golden and crisp. Remove from the oven and let cool.

Top each baguette slice with a generous spoonful of the chopped liver, arrange on a platter, and serve.

CHICKEN-FRIED QUAIL

Delectable, crispy quail make an elegant alternative to traditional fried chicken. Unlike chicken, quail can be fried whole quickly, evenly, and with minimal fuss. For an extra-spicy version of this dish, fry the quail in the chilli-laden lard left over from making *ciccioli* (page 245). **SERVES 4**

8 semi-boneless quail (see page 67)
360g plain flour
1 tablespoon fine sea salt
1 teaspoon freshly ground black pepper
1 tablespoon paprika
2 teaspoons ground cayenne pepper
1 litre lard
Coarse sea salt, such as fleur de sel or
    Maldon, for sprinkling

BUTTERMILK BRINE
40g fine sea salt
40g honey
120ml boiling water
720ml buttermilk

To make the brine, combine the salt and honey in a heatproof container or bowl large enough to hold the quail. Pour the boiling water over the salt and honey and stir to dissolve. (Be sure to measure the water after it has come to a boil, as some will evaporate as it heats and measuring beforehand will result in an overly salty brine.) Let cool to room temperature, then stir in the buttermilk.

Pat the quail dry and fold the wing tips behind their backs. Submerge the birds in the buttermilk brine, topping them with a plate if necessary, and leave at room temperature for 2 hours.

Meanwhile, combine the flour, fine sea salt, black pepper, paprika, and cayenne in a wide, shallow bowl.

Remove the quail from the brine, drain them well and pat them dry. Dredge each quail in the flour mixture. In a large, deep cast-iron skillet or frying pan heat the oil until it reaches 180°C (gas 4). Shake the excess flour from each quail, then fry them, breast side down, making sure not to crowd the pan. Cook for about 4 minutes on each side, reducing the heat if they're browning too quickly. Use a slotted spoon or meat fork to turn them; tongs will break off the delicious crispy skin that some say is the best part.

Transfer the birds to a plate covered with paper towels to drain and sprinkle with coarse sea salt. Make sure the finished birds lie in a single, uncrowded layer so that they stay crispy.

Rabbit

Rabbit is a delicious, nutritious, often overlooked choice for those in search of a more sustainable meat option. All bunny jokes aside, rabbits really do reproduce quickly and abundantly. Their meat has a delicate flavour that adapts well to a variety of recipes.

Rabbits are sold skinned and the meat should be a lovely pink. They are usually sorted according to size. Young, tender rabbits that weigh 1.4–1.8kg are perfect for sautéing. Birds weighing 1.8–3kg, with firmer and somewhat less tender flesh, are ideal for roasting. Older rabbits weighing over 3kg are ideal for sausages, rillettes (page 244), and terrines. A rabbit is often sold with its liver and kidneys, which can be grilled (as in the rabbit *spiedini* on page 165) or added to sausages and terrines.

As with poultry, source quality rabbit from butchers that have a high turnover or will special-order fresh rabbit if they do not regularly have it in stock. Rabbit should generally be consumed within a week of processing.

Butchering Rabbit

Rabbit is a great place to begin if you're interested in learning how to butcher larger beasts such as lambs, goats, and pigs. The skeletal and muscular structures are fairly similar, but the rabbit is obviously much smaller and more manageable.

Rabbit on the Bone

Use the method illustrated below to cut a rabbit into six pieces for braising, grilling, or frying.

Boned Rabbit

To separate all of the meat from the rabbit carcass for pâté or sausage, use the method illustrated on pages 86–8. In addition to the meat, you will yield the carcass, forelegs, and offal.

Whole Boned Rabbit

To completely debone a whole rabbit while leaving it intact (as for Rabbit Porchetta, page 177), use the method illustrated on pages 89–90.

RABBIT ON THE BONE

◀ **1. Harvest the offal.** Lay the rabbit on its back on a cutting board and check its abdominal cavity for the kidneys and liver (they are often left attached). If they are present, remove them with a sharp knife.

▶ **2. Cut the rabbit into thirds.** Leaving the rabbit on its back, remove the front legs (shoulders) by cutting through the backbone with a cleaver, directly behind the last rib. Cut the remaining two-thirds of the rabbit in half by cutting through the backbone with a cleaver, 2.5cm above where the leg meets the loin.

3. Cut the rabbit into sixths. Using a cleaver, split each section lengthwise through the backbone to create six roughly equal-sized pieces.

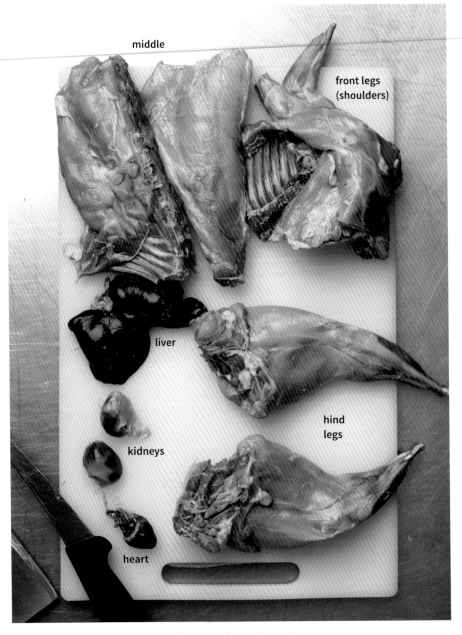

4. The result: Offal, front and hind legs, and middle sections.

1. Harvest the offal and tenders. Lay the rabbit on its back on a cutting board and check its abdominal cavity for the kidneys and liver (they are often left attached; see photo for step 1 on page 84). If they are present, remove them with a sharp knife. Cut the tenders from the inside of the cavity (they are located just underneath where the liver was and will be about 4cm long, nestled into the backbone).

2. Remove the front legs (shoulders). With your hands, gently pry the shoulders away from the body. Using a boning knife and following the natural seam under a front leg, cut away the leg until it comes away entirely. Repeat with the second leg. Because the front legs contain only a tiny amount of meat, it is usually not worth the effort to bone them. Instead, set them aside for a stew, stock, or rillettes.

3. Remove the hind legs. With your fingers, feel where the leg bone connects to the back. Then, positioning the knife about 2.5cm above where the two bones meet, cut straight through the meat to the back. Hold the rabbit down on the board and bend the leg to free the ball socket of the leg from the backbone. Continue to cut along the edge of the backbone until you've freed the leg completely. Do the same with the other leg.

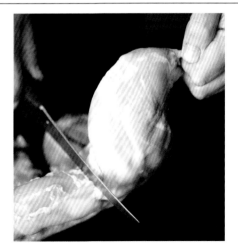

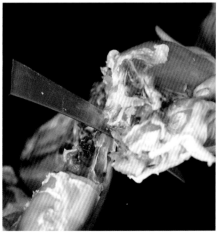

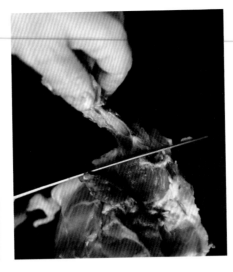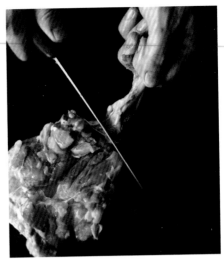

4. Separate the leg meat from the bone. Starting at the ball joint where you removed the leg from the backbone, cut down one side of the leg bone, working right next to the bone, until you come to the end. Next, go down the other side of the leg bone all the way to the end. Pull the ball joint away from the leg and remove the underside of the leg muscle from the bone. Repeat with the other leg.

5. Remove the loins and saddle. 'Saddle' is the term used to describe the meat attached to the ribs below the loin on a rabbit (analogous to the belly or breast meat of pork or chicken). Starting at the neck end, use your boning knife to cut along one side of the backbone, gently peeling the meat away from the bone with your fingers as you cut. Peel all of the meat off the side of the rabbit, all the way down to the bottom of the ribs and leaving as little meat behind as possible. Repeat on the other side.

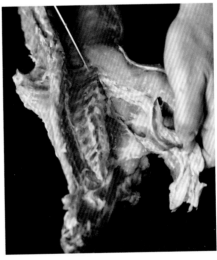

6. Separate the loin from the saddle. Using your knife, cut the flat rectangular saddle away from the cylindrical loin muscle.

CONTINUED

IN THE BUTCHER'S

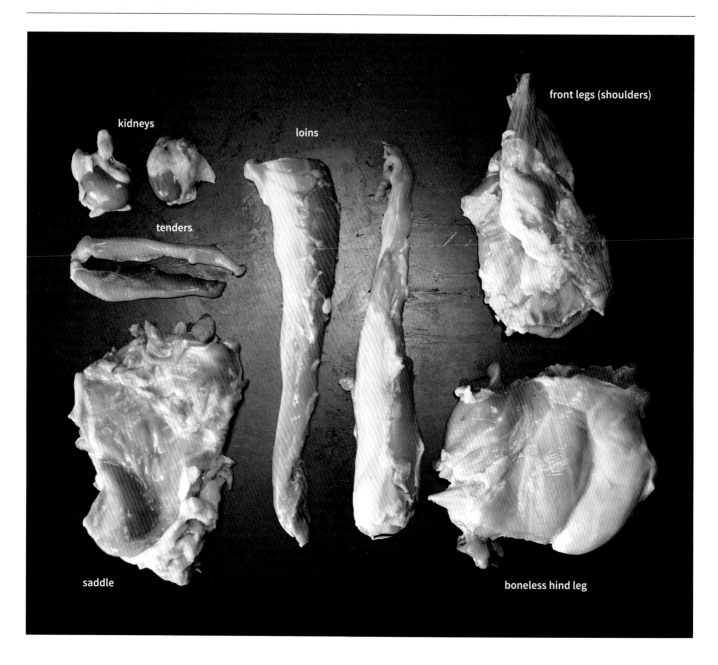

IN THE CHARCUTERIE

1. Harvest the offal and tenders and remove the front legs. Follow steps 1 and 2 for Boned Rabbit (see page 86).

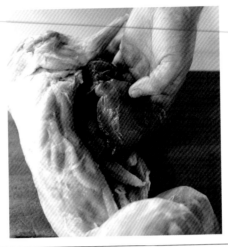
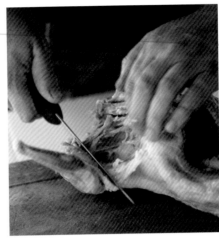

2. Cut the meat away from the ribs. Starting at the neck end, and working with the tip of the boning knife pressed against the rib cage, cut the meat away from the rib bones. When you reach the fleshy loin area, which is attached to the backbone, curve the tip of your knife around the loin until you hit the backbone. Repeat this action down the length of the rib cage on each side of the rabbit until you reach the hind legs.

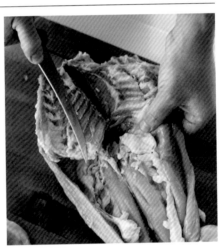

3. Remove the backbone. With your boning knife, sever the tip of each vertebra while pulling the backbone away from the meat with your other hand. Take care not to pierce through the meat.

CONTINUED

4. Remove the hind leg bones. To separate each leg bone from its hip socket, push the leg backward until you feel the ball joint at the top of the leg pop out of its socket.

5. Carve the hip socket out of the meat, then slice along each side of the leg bone, pressing against the bone as you go.

6. Slip the knife underneath the leg bone to free it from the muscle, scraping any extra meat away from it as you go. Repeat with the other hind leg.

IN THE CHARCUTERIE

Rabbit Porchetta,
page 177

RABBIT CACCIATORE

This rustic, easy-to-prepare stew uses the whole rabbit. Although rabbit is quite lean, braising the meat on the bone helps to keep it moist and delicious. If your rabbit comes with its liver and kidneys, these can be chopped and added along with the vegetables to produce an even earthier stew. Start preparing your rabbit the night before you plan to cook it. Serve the rabbit and its sauce over creamy polenta or buttered egg noodles. **SERVES 6**

1 rabbit, 2–2.3kg, cut into 6 pieces

Fine sea salt and freshly ground pepper

3 tablespoons olive oil

6 cloves garlic, lightly crushed

3–4 sprigs mixed thyme and savory, plus
 1 tablespoon finely chopped fresh
 savory and 1 tablespoon finely chopped
 fresh thyme

70g diced bacon, guanciale, or pancetta,
 home-made (page 299, 296,
 or 295) or shop-bought

170g minced yellow onion

55g finely chopped celery

70g peeled and finely chopped carrot

300g diced red or yellow sweet pepper

30g dried porcini mushrooms, rehydrated
 (see page 19, Dried Wild Mushrooms)
 and coarsely chopped

120ml dry white wine

250g tomato purée

480ml rabbit, game, or chicken broth (see
 Basic Rich Broth, page 44)

1 or 2 dried cayenne peppers or other hot
 chillies

The day before you want to cook your rabbit, generously season it with salt and pepper and place it in a single layer in a large, shallow bowl or baking dish. Add the olive oil, garlic, thyme and savory sprigs, and toss to coat the rabbit pieces evenly. Cover and refrigerate overnight.

The following day, remove the rabbit from the fridge and bring to room temperature. Remove the rabbit pieces and garlic cloves from the marinade and reserve separately. Discard the remaining marinade.

Heat a large, deep frying or sauté pan over a medium heat. Add the bacon and let it brown slowly, releasing its fat. When most of the fat has rendered, using a slotted spoon, transfer the bacon to a small plate and reserve.

Add the rabbit pieces to the pan and brown well on all sides. Using tongs or a slotted spoon, transfer the rabbit pieces to a large plate and set aside. Add the onion, celery, carrot, sweet pepper, porcini, and reserved garlic cloves to the pan. Sauté for about 10 minutes, then add the wine, tomato purée, and broth, bring to a simmer, and cook for 5 minutes more.

Return the rabbit to the pan and add the reserved bacon, chillies, and chopped savory and thyme. Stir to combine, then turn down the heat to low, cover, and cook for about 1 hour, until the meat is tender and beginning to loosen from the bone. Taste for seasoning, then serve.

Pork, the Provider

No other animal is as essential to the charcuterie as the pig. It provides us with the makings for sausage, ham, salami, bacon, and terrines; gives us succulent cuts suitable for roasting, smoking, grilling, and braising; and delivers tasty offal and skin, an abundance of bones for rich, gelatinous broths, and ample lard for baking, frying, and confits. The pig pretty much has it all.

All domesticated pigs originated from the wild boar that roamed the forests and swamps of Asia and Europe forty million years ago. Conflicting claims exist on the first official domestication of swine, which occurred some time around or before 5000BCE, in the Middle East, the eastern Mediterranean, or possibly China. Regardless of who planted their flag in Swineville first, it was the Romans who tinkered with hog breeding and spread pork production throughout their empire. When Christopher Columbus set sail on his second voyage to the New World in 1493, eight pigs were among his passengers.

From a farming perspective, pigs are brilliant. Female pigs, or sows, give birth to an average litter of ten piglets after a gestation period of only four months and can produce several litters annually. Piglets grow fast and can increase their weight by 5,000 per cent in just six months. For farmers, this means a higher return for time invested than for any other domesticated animal. Pigs can also eat almost anything people can eat. Their omnivorous nature allows them to dine on everything from fallen acorns, foraged roots in the forest, and surplus crops of sweet potatoes, coconuts, barley, or corn to kitchen scraps on the small family farm, converting waste and leftovers into delectable meat.

At the Pork Store

From its flavourful head to its springy tail, pork is good food. Nearly every part of the pig can be eaten or used in the charcuterie, including its liver, ears, feet, tongue, skin, and blood.

When purchasing pork, the best indicators of quality are the colour and texture of the meat and fat. Despite the 'other white meat' campaign, pork is actually red meat and the most desirable colour for the meat is deep pink to nearly red, with a fair amount of intramuscular fat or marbling dispersed throughout. The fat should be almost bright white. Avoid pork that has a grey or brown cast, a coloration caused by oxidation, which can indicate a lack of freshness. Both the meat and fat should have a firm texture. Cuts for roasting or smoking, such as hams, shoulders, loins, and bellies, should have a fat cap that is at least 1cm thick, and cuts for charcuterie should generally have an even thicker cap.

Happy pigs taste better. Look for meat from pigs that have been pasture-raised or raised without confinement. By living in a natural way, they get exercise that builds up their muscles and produces meat with a deeper, porkier flavour. Pigs that spend a good amount of time outdoors also tend to develop more and better subcutaneous fat as a natural defence against cold weather.

Heritage Pigs

Heritage pork breeds are more than just fancy names and trumped-up pedigrees. They are distinguished breeds with unique genetics that contribute to the quality and flavour of their meat and fat. Modern industrial breeds have too little fat and not enough flavour to be useful in the charcuterie, so we rely solely on heritage breeds to help us achieve a robust, old-world flavour. Heritage breeds are well suited to a life spent in the great outdoors and are generally raised on small family farms where they have room to roam and time to grow slowly, developing rich flavour, good marbling, and back fat, which make for better roasts, sausage, and cured meats.

Berkshire: Black-and-white hogs prised for their sweet, nutty flavour, Berkshires are probably the most popular of the heritage breeds today. They have well placed intramuscular fat, or marbling; bright pink meat; and a fat cap that is a good thickness for roasts and chops. We especially love Berkshire shoulder roasts with their perfect balance of meat and fat, great for slow roasting, pot roasting, or braising. In the charcuterie, the all-purpose Berkshire accounts for the majority of the pork we use to make bacon, ham, and *salumi*.

Duroc: The Duroc is a muscular, red-coated American pig. Durocs tend to be a bit leaner than other heritage breeds, but their ample meat is well marbled and they produce juicy roasts and chops. Some of the best pork we get at the Fatted Calf is milk-fed Duroc from Good Farms in Olsburg, Kansas. The Goods raise a small portion of their glorious Durocs on a diet of both pasture and raw fresh milk, and the result is extremely tender, sweet meat.

Gloucestershire Old Spot: This white-coated hog with black spots and large, floppy ears that curl over its head is shy and sweetly tempered. The Old Spot was nicknamed the Orchard Pig because it was often grazed in pear and apple orchards, where it fed on windfall fruit. Although the meat of the Old Spot is

naturally a paler pink than the meat of Tamworth hogs, it is flavourful and rich, with abundant marbling and a thick fat cap. The plentiful fat makes this breed a great choice for most charcuterie, particularly for *lardo* and hams.

Large Black: As its name conveys, the Large Black is a black big-bodied hog that can reach upward of 300 kilos when fully mature. The breed, which is thought to have origins in China but was refined in England, has deep reddish meat and a rich, earthy flavour. It also has a generous covering of fat, making it a good choice for *salumi* production. At the Fatted Calf, we get a barley-fed Large Black hog from a local farm about once a month that we use to generate *lardo*, pancetta, *coppa*, and salami.

Mangalitsa: Originally from Hungary, the Mangalitsa, sometimes called the Woolly Pig, has a thick, woolly coat best suited to cold northern climates. The Mangalitsa is a very slow-growing lard pig and has meat with nearly double the marbling of average pork and an abundance of thick back fat. Probably because they are so unusual and slow growing, Mangalitsas also generally fetch double the price of your average heritage pig.

Tamworth: The Tamworth is a muscular, red English hog with some genetic roots in Ireland. It is blessed with a long middle and ample belly and is thought to be the best of the bacon breeds. It has a pronounced nutty flavour that is prized in the charcuterie for *salumi* and prosciutto.

The Whole Beast

A whole pig is an impressive beast. Even a smaller pig, or roaster, weighs in at roughly 25–45kg. Optimal market hogs generally weigh between 45 and 90kg and are 6–8 months old. Larger, more mature pigs, those over 90 kilos, have more fat, larger muscles, and a more intense flavour, making them perfect for use in the charcuterie. When purchasing a whole pig, try to arrange well in advance with your butcher or a local farmer. Be sure to request the head, which can be used for Pig Head Pozole (page 126), Guanciale (page 296), or various terrines, and the offal, as these are sometimes removed at the time of slaughter. If you plan on butchering the pig yourself, it is highly recommended to have the pig split in half to make the initial cutting more manageable.

Butchering a Whole Pig into Primals

Pigs have three main parts, or primals, from which most other marketable cuts are produced. They are the shoulder (see pages 112–15), the middle (which includes the belly and the loin), and the ham (see pages 120–21). And every whole beast is blessed with two of each! For illustrated instructions on how to butcher a pig into primals, refer to pages 99–100.

The Middle: Loin and Belly

The pork middle is just that: the middle piece between the shoulder, or front end, and the hind leg. The middle is generally broken down into smaller, more manageable parts – although preparations exist that utilise the whole middle (see Whole Boneless Middle, right). Depending on how you choose to butcher it, the middle can yield rib and porterhouse chops, loin back ribs, belly, spare ribs, tenderloin, and bone-in and boneless loin roasts. See page 101 for illustrated instructions.

The Belly

A bone-in, skin-on whole belly makes a fabulous, rich roast or succulent braise. Or, if you want a boneless belly, you can separate the spare ribs. See page 102 for illustrated instructions. The belly can also be skinned (see Skinning the Pig, page 124) if you plan to make bacon (page 299).

The Primal Loin

The pig's primal loin is an excellent roasting cut. Because it is too large for everyday cooking, it is generally broken down into more practical bone-in or boneless roasts, chops, and ribs. (The number of ribs varies according to breed.) The first step is to divide the primal loin into the rack and loin. The loin is then cut to yield the tenderloin and a boneless roast. Next, you must choose between cutting the rack into bone-in rack roasts and chops or a boneless loin roast and loin back ribs. Instructions for all of these options are found on pages 103–105. If you want skinless roasts and chops, skin the loin prior to beginning to cut (see Skinning the Pig, page 124).

Whole Boneless Middle

A whole boneless middle (meaning both the belly and the primal loin are kept together) makes an impressive roast. We use this butchering technique for the Cuban (page 181) but it can also be used for Italy's famous *arista alla porchetta*. See pages 106–7 for illustrated instructions.

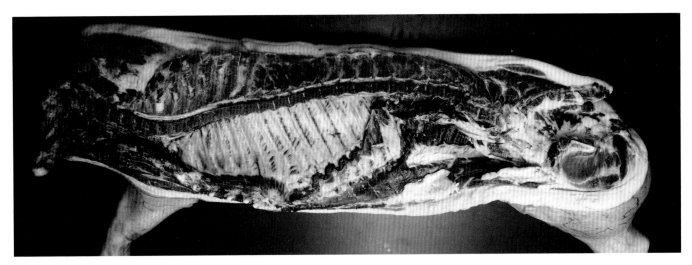

1. Split the pig into halves. If the head is attached, saw if off where it connects to the neck. To split the pig, saw directly through the backbone beginning from the tail end and continue all the way up through the neck. If the pig is hanging, this is made easier if a second person holds the pig in place.

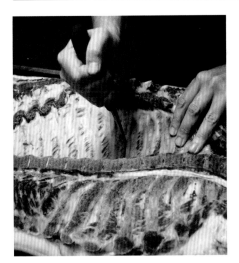 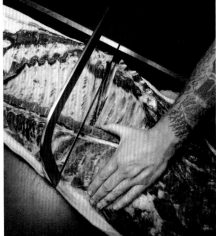 

2. Separate the shoulder. Place a half cut side up. Make a tiny mark with a boning knife between the fourth and fifth ribs. Using the mark as a guide, push the tip of a boning knife all the way through the meat to the other side of the rib cage and then through the skin. Now, make one long, smooth cut away from the backbone through the bottom of the rib cage. (If you have difficulty cutting through the bottom of the rib cage with a boning knife, use the saw to cut through the bones, then resume using the knife.) Turn the knife around and follow the initial cut all the way down to the backbone. Saw through the bone, then use a chef's knife or scimitar to remove the whole shoulder in one clean cut. The entire front piece of shoulder should now be completely separated from the rest of the pig.

CONTINUED

3. Separate the leg from the middle. With the half again cut side up, and looking at the hind leg, find the area on the backbone where the leg starts to curve into the tail. Make a perpendicular cut through the belly below the curve of the backbone with your knife, then turn the knife around and follow the cut all the way back up to meet the backbone. Saw through the backbone. Line up your knife with the initial cut and cut all the way through the leg, separating it from the middle.

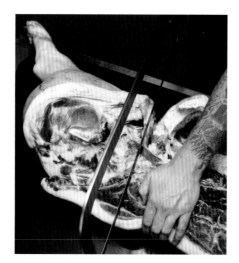
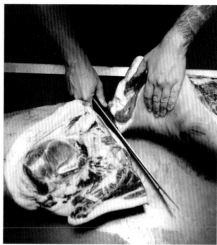

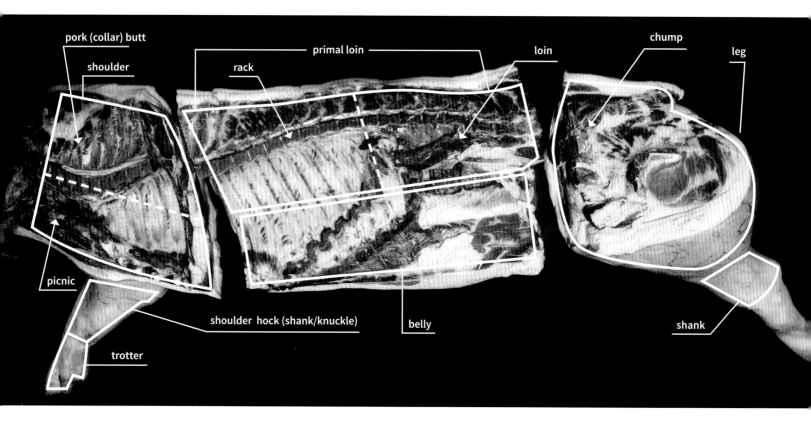

pork (collar) butt

shoulder

primal loin

rack

loin

chump

leg

picnic

shoulder hock (shank/knuckle)

belly

shank

trotter

THE MIDDLE: PRIMAL LOIN AND BELLY

1. To separate the belly from the primal loin, score the meat where you plan to cut. Place the middle on a cutting board cut side up. Starting at the shoulder end, measure about 12cm from the backbone to the belly. Mark the spot with your boning knife. Without cutting too deeply into the meat, score the meat and ribs from your mark at the shoulder end all the way to the opposite end of the middle in a straight line.

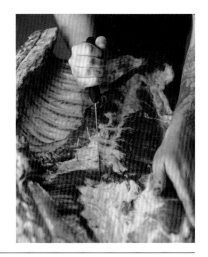

2. Cut through the ribs. Using your saw, cut through each rib, one at a time, by sliding your hand underneath the middle to elevate each rib as you are cutting it. Begin sawing at the shoulder end and continue all the way to the last rib. (Although this can be a little unnerving at first, know that you would need to keep going for quite a bit to cut all the way through the meat and skin to get to your fingers. All that's getting cut at this point are the ribs.)

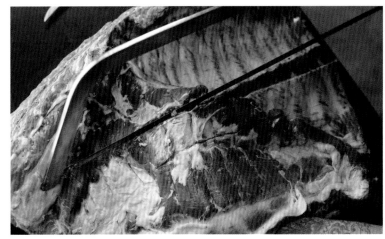

3. Completely separate the belly from the primal loin. Once all the ribs have been split, using your boning knife, follow your initial scoring mark to cut all the way through the skin to separate the belly from the loin completely.

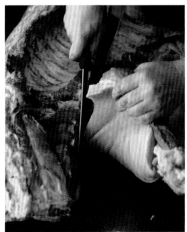

Remove the spare ribs from the belly. Begin at the shoulder end where the rib sits flush against the edge of the belly. Using a boning knife, and cutting parallel to the work surface, carve the ribs away from the belly. If you will be smoking, roasting, or braising the rack of ribs, leave about 1cm meat attached to the underside. If the ribs are destined for the stockpot, leave as much meat attached to the belly as possible.

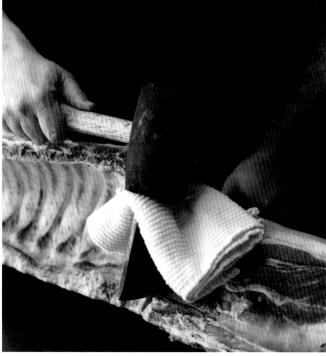

1. Split the primal loin into the rack and loin. Using a chef's knife or scimitar, make a cut behind the last rib of the primal loin, nearest where the leg was removed. Switch to a meat cleaver. Nestle the cleaver in the cut, against the backbone. Lay a folded towel on top of the cleaver. Holding the handle of the cleaver in one hand, use your other hand to tap the back of the cleaver with a rubber mallet until the blade goes through the bone completely. You now have a rack (longer section) and loin (shorter cut).

2. Bone the loin to yield the tenderloin and a boneless roast. The tenderloin, or fillet, is the small cylindrical muscle running along one side of the loin. Using your hand and a boning knife, peel the tenderloin away from the bone to separate it completely.

CONTINUED

3. Bone the remainder of the loin to create a boneless loin roast. Position the knife directly against the bone and cut between the bone and the meat to separate it to create a boneless roast. Save the bones for broth.

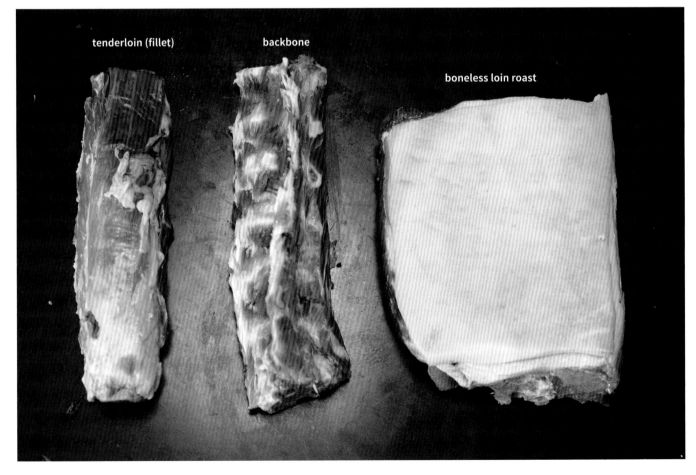

tenderloin (fillet)   backbone

boneless loin roast

4a. Cut rib chops and roasts from the rack. (Alternatively, for loin back ribs, proceed to step 4b.) Rack roasts and rib chops are cut using the same technique. Using a chef's knife or scimitar, cut between two ribs (to cut chops, cut between the first two ribs; for roasts, count the number of ribs you would like in your roast and make the cut just after) through the loin meat to the backbone to make an initial cut. Switch to your meat cleaver. As in step 1 on page 103, nestle the cleaver in the cut, against the backbone. Lay a folded towel on top of the cleaver. Holding the handle of the cleaver in one hand, use your other hand to tap the back of the cleaver with a rubber mallet until the blade goes through the bone completely. Repeat this for as many chops or roasts as you want to generate.

4b. Separate the loin from the loin back ribs. Stand the rack on your cutting board, with the sawed end of the ribs facing up. Using your boning knife and working closely against the bone, cut away the loin muscle to separate it from the back ribs.

WHOLE BONELESS MIDDLE

1. Start with a whole pork middle, preferably with the backbone removed (as in the photo at right; you can ask your butcher to do it, or remove the backbone yourself in step 5). Cut through the ribs as if you were going to separate the belly from the primal loin (see steps 1 and 2 on page 101). Do not cut all the way through the meat and skin, though.

2. Remove the ribs. Place the middle on a cutting board on its side with the backbone facing you. Remove the ribs one-by-one: using the tip of a boning knife, cut down one side of the rib closest to you, cutting only as deep as the bone. Work your way around the tip of the rib, then cut down the other side until you reach the backbone, essentially outlining the rib with your knife. Using your hands, wriggle the end of the rib free, then cut the meat away from the bone as you pull the rib away from the belly. When most of the rib bone is free of the meat, press the rib down and twist it to release the bone from the backbone. Carve away the remaining meat from the joint. Repeat the process to remove the remaining ribs.

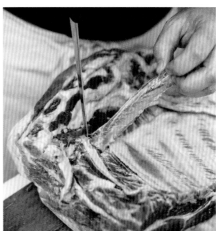

3. Remove the cartilage. Cartilage runs along the centre of the middle where the ribs connected to the breastplate; locate each piece with your fingers, then, using your boning knife, cut it away from the meat.

4. Check for pieces of the shoulder blade. A piece of bone from the shoulder blade may be nestled between the meat and skin on the shoulder side; if so, cut it away using your boning knife.

5. Remove the backbone, which extends from the shoulder end to the leg end. If your middle still has a backbone, remove it by peeling the tenderloin away from the bone, being careful to leave it attached to the muscle just above it. Directly underneath the tenderloin is the finger bone, which is the bone that makes the long vertical line in a T-bone. Working directly against the bone, slip your boning knife behind each of these finger bones, cutting down to the backbone. Once all of the finger bones have been exposed, cut the backbone away from the loin in one piece, working from the shoulder end to the leg end.

▶ **6. Score the skin.** Turn the now-boneless middle over so that the skin side is facing up. Using the tip of your knife and a light touch, scratch the surface of the skin, making straight, parallel lines, 5cm apart, that stretch from the shoulder end to the leg end.

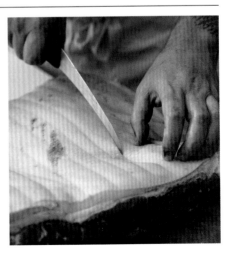

FIVE-SPICE LOIN BACK RIBS

There are two types of pork ribs: spare ribs, the long ribs that rest on the inside of the belly, and loin backs, the smaller ribs that sit alongside the loin. Spare ribs are big, tough, and meaty, and benefit from a long, slow roast in the smoker. But tender little loin backs can be ready in a fraction of the time and can even be roasted in the oven with great results.

These ribs make great hors d'oeuvres or can be served as part of a larger meal along with Duck and Lemongrass Sausage Patties (page 204), steamed rice, and vegetables. **SERVES 4**

2 racks loin back ribs

1 teaspoon fine sea salt

2 teaspoons sugar

2 teaspoons Five Spice (page 11)

3 tablespoons dark soy sauce

4 teaspoons dry sherry

6 cloves garlic, minced

Loin back ribs have a membrane on the underside that must be peeled away prior to cooking. To loosen the membrane, score the underside of the first rib bone lengthwise on each rack. A small piece of the membrane will separate. Grab hold of this piece with a clean, dry towel and peel away the entire membrane.

In a small bowl, stir together the salt, sugar, Five Spice, soy, sherry, and garlic. Place the racks in a large, shallow bowl or baking dish and cover with the marinade, turning to coat. Cover and refrigerate for at least 2 hours or up to overnight.

To roast the ribs in the oven, preheat the oven to 180°C (gas 4).

Take the ribs, reserving any extra marinade, and place them, meat side up, on the rack of a roasting tin. Roast for 30 minutes, then baste with the reserved marinade and continue to roast for about 30 minutes longer, until the meat begins to ease back from the ends of the bones.

To grill on the barbecue, prepare a medium fire and place the ribs on the grill grate directly over the coals. Alternatively set your oven grill to medium. Grill, turning and basting the ribs with the reserved marinade every 10 minutes, for about 1 hour, until the meat begins to ease back from the ends of the bones.

Let the ribs rest for 5 minutes before cutting. To serve, slice between the bones to separate into individual ribs.

CHERMOULA-MARINATED PORK CHOPS

Chermoula is a North African marinade traditionally used to flavour fish and vegetables. In an unconventional use, we also like to slather this zesty concoction on pork chops or roasts. Grating the garlic, onion, ginger, and turmeric on a fine-rasp Microplane grater or the smallest holes of a box grater intensifies their flavour and helps to tenderise the meat. You can also double the marinade and use the extra to season roasted or grilled vegetables to serve alongside your chops.

SERVES 4

4 pork rib chops, 225–280g
Fine sea salt

CHERMOULA
1¹/₂ teaspoons cumin seeds, toasted
    (see page 11)
¹/₂ teaspoon coriander seeds, toasted
    (see page 11)
¹/₂ teaspoon peppercorns
¹/₄ teaspoon chilli flakes
2 cloves garlic, grated
50g grated onion
¹/₂ teaspoon peeled and grated fresh ginger
1 teaspoon peeled and grated fresh turmeric,
    or ¹/₂ teaspoon ground turmeric
2 teaspoons unsmoked Spanish paprika
Grated zest of ¹/₂ lemon
2 tablespoons finely chopped fresh flat-leaf
    parsley
2 tablespoons finely chopped fresh
    coriander
3 tablespoons olive oil

Salt the pork chops on both sides and the rind of fat on the outer rim of each chop.

To make the *chermoula*, in a spice grinder, combine the cumin, coriander, peppercorns, and chilli flakes and grind finely. Transfer the ground spices to a large, shallow bowl, add the garlic, onion, ginger, turmeric, paprika, lemon zest, parsley, coriander leaves, and 2 tablespoons of the olive oil, and stir to combine.

Add the salted chops to the bowl and roll them in the marinade to coat evenly. Wrap the marinated pork chops tightly in cling film and refrigerate overnight.

Heat a large cast-iron or heavy-bottomed frying pan over a medium heat and add the remaining oil to the pan. When the oil is hot, add the chops in a single, uncrowded layer and cook, turning once, for about 5 minutes on each side, until nicely browned but still tender and juicy inside. Serve at once.

The Shoulder

The shoulder is probably the most versatile, delicious cut of any animal and the one used more than any other in the charcuterie. It is excellent for brochettes, sausage, salami, and terrines and can be slow roasted, braised, smoked, or cured. This primal cut is comprised of the pig's front leg and shoulder, which yields the trotter (foot), the shoulder hock, the picnic, and the pork (collar) butt. Once you have severed the shoulder from the remainder of the pig half (see page 99, Butchering a Whole Pig into Primals), breaking it down into the smaller cuts is simple. See below.

THE SHOULDER

1. Saw the trotter away from the shoulder hock. With the shoulder skin side down on a cutting board, bend the foot back and forth to locate the joint. Holding the foot with one hand, saw completely through the bone to separate. The trotter can be set aside for broth or Head Cheese (page 250).

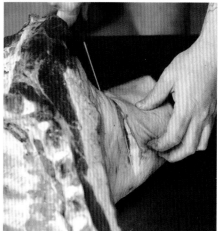

2. Remove the foreshank in two sections. Locate the elbowlike bone that comes out at an angle from just below the ribs. Measure halfway between that bone and the end where the trotter was removed. Using a boning knife, and working parallel to the cut you made to remove the trotter, slice through the skin and meat down to the bone. Use the saw to cut through the arm bone, then resume cutting the meat and skin with the knife to separate completely. To remove the second section, make the same cut directly underneath where the elbow bone bends. The foreshanks can be used for Braised Ham Hocks (page 285) or Classic Cassoulet (page 57).

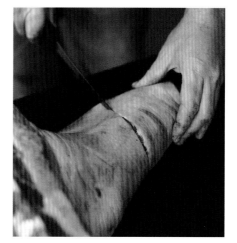

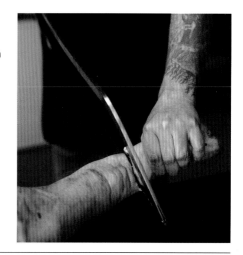

3. Separate the picnic from the pork (collar) butt. Measure 10cm from the backbone towards the foreshank end. Using the saw, cut through the ribs parallel to the straight edge of the shoulder. Cut through the rest of the shoulder with your knife, sawing through the shoulder blade if necessary. You will have a squared-off piece with the ribs attached to the backbone, the pork butt, and a more triangular and irregularly shaped piece, the picnic.

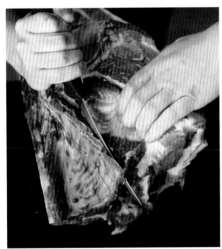

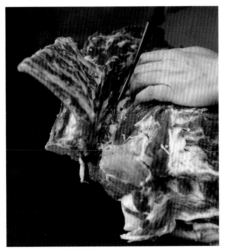

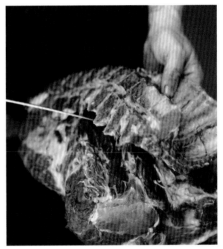

4. Bone the pork butt. The square-cut pork butt rests on rib bones that attach to both the backbone and the shoulder blade. Place the cut rib side up on the cutting board. To remove the ribs and backbone, work your boning knife directly behind the four ribs that attach to the backbone. Cut along the length of the ribs to the backbone, then continue cutting along the backbone until all of the bones separate from the shoulder.

CONTINUED

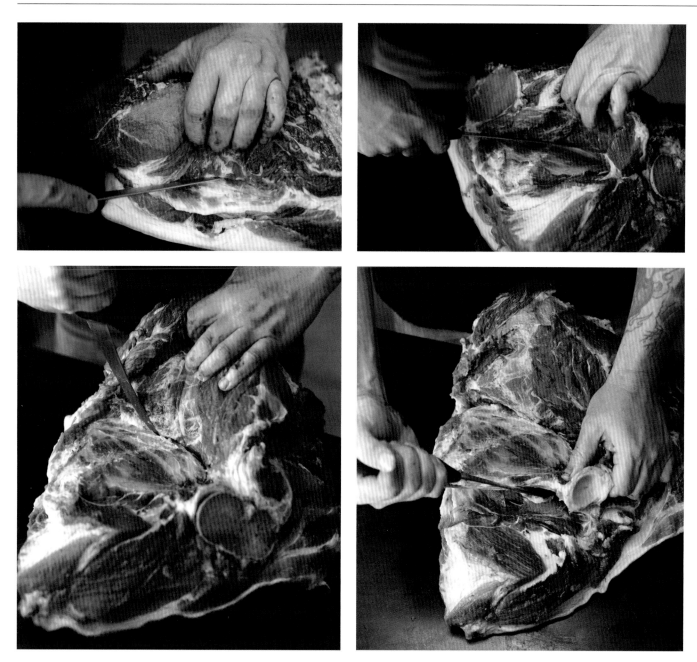

5. Remove the shoulder blade. Turn the pork butt skin side down on the cutting board. Locate the edge of the shoulder blade, which is nestled between the skin and the spot where the ribs were removed. Insert your boning knife directly above and parallel to the bone, keeping the flat of the knife against the shoulder blade. Continue to cut until the shoulder nearly opens like a book, exposing the whole shoulder blade. Carve down either side of the shoulder blade with your knife to outline it, then cut underneath the shoulder blade to free it.

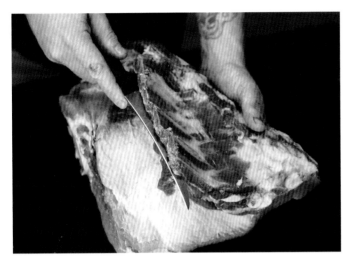

6. Bone the picnic. The picnic contains a piece of the arm bone and four small pieces of rib bone. First, using the boning knife, carve the rib bones away from the rest of the picnic. To remove the arm bone, butterfly the meat on either side of the bone where you removed the foreshanks. Working against the bone, cut around the end of the arm bone, then cut around the underside to remove the bone completely.

CONSIDER THE BLADE ROAST

The blade is a roast taken from the pork butt section that includes the first four ribs of the pig and all the muscle in between those ribs and the shoulder blade. It is the holy meeting place between the loin and the succulent shoulder. It is a great little roast that can easily be seasoned or stuffed and is perfect for feeding three to four people. Once you remove the blade, the remainder of the pork butt can be cut into smaller shoulder roasts or cubed for sausage.

To remove this roast, turn the cut side towards you and locate the shoulder blade. You will be removing the piece between the ribs and the inside of the shoulder blade. Turn the shoulder skin side down and rotate it so the cut end of the ribs is facing away from you. Saw through the backbone directly before the first rib. Cut through the shoulder with a chef's knife or scimitar until you hit the shoulder blade. Saw through the shoulder blade, then return to the knife and cut through the rest of the shoulder. Next, place the knife at the top of the ribs and carve towards the shoulder blade. Once the knife gets to the top of the shoulder blade, press your knife towards the bone and continue cutting against the bone, following it all the way down to the bottom. When you reach the bottom of the shoulder blade, turn your knife at a right angle towards the skin of the shoulder and cut through the skin. You now have your blade roast, excellent for stuffing and slow roasting, as well as for smoking or braising.

PORK BOLLITO MISTO

Bollito misto is a northern Italian speciality that calls for simmering a variety of meats together in a rich broth. It's a classic Sunday supper dish prepared by *Nonna*. The broth is usually served as a first course and the meats are sliced and served as a second course accompanied with a few sauces and some simply prepared vegetables. What belongs in the broth is a subject of great regional pride and debate. In some elaborate versions, beef, veal, pork, chicken, tongue, and *cotechino* (page 223) or *zampone* (a stuffed pig's trotter) are stewed together and served with seven sauces.

In this simplified pork version, the mix consists of belly, tongue, and shoulder. You can add different cuts or meats, but make sure you choose tougher cuts, such as neck, hock, and shoulder, that are more suitable for braising. As with all braises, making *bollito misto* a day or two ahead of time allows the flavours to bloom and makes degreasing the broth and slicing the meats much easier. **SERVES 6**

900g boneless pork butt or picnic in one piece, preferably with skin

900g boneless pork belly in one piece, preferably with skin

3 pork tongues

Fine sea salt

1 teaspoon freshly ground pepper

3/4 teaspoon chilli flakes

2 allspice berries, ground

Pinch of freshly grated nutmeg

3 litres pork or chicken broth (see Basic Rich Broth, page 44), or a mixture

360ml dry white wine

1 yellow onion, peeled but left whole

4 cloves garlic, peeled but left whole

1 small dried bay leaf

6 slices country-style bread, toasted and rubbed with garlic (optional)

Horseradish Salsa Verde (page 323) or Cherry Mostarda (page 320), for serving

Season the pork butt, pork belly, and tongues on all sides with salt. Place in a large, shallow bowl or a large baking dish, cover, and refrigerate overnight.

The next day, sprinkle the pepper, chilli flakes, allspice, and nutmeg over the meaty side of the pork belly. Roll up the pork belly lengthwise and tie it with butcher's twine as you would a roast, tying it as tightly as possible (see page 168).

Place the pork butt, belly, tongues, broth, wine, onion, garlic, and bay in a tall, narrow stockpot (12 litres is perfect). Bring to a simmer over a medium heat, skimming the surface as you would when making broth. Cover partially and cook for 2 hours, then check each piece. They should all cook at about the same rate, but some might be done sooner than others depending on thickness. If any of the meat starts to look like it is about to start falling apart, use a slotted spoon to transfer it to a plate. If the level of the broth drops more than 2.5cm or so below the tops of the meats, add a little more broth or some water. Taste the broth and season if necessary; it should be delicious all on its own, as the salted meats and the spices on the pork belly will impart a little seasoning, but occasionally it will need a little nudge.

After 2 hours at a good simmer, the meats should be fully cooked and well softened. Slice off a small piece of the pork butt and taste it. It should still have a solid texture but not be rubbery. Put all of the meats back in the pot and let cool to room temperature, then cover and refrigerate overnight.

The next day, take all of the meats and the onion out of the chilled broth. Discard the onion and any cold fat on the top of the broth. Remove the string from the belly but do not unroll it. Slice the pork butt and belly into about 0.5cm slices, carving against the grain. Slice the tongues 1cm thick across the grain.

CONTINUED

PORK BOLLITO MISTO, continued

Return all of the meats to the pot.

Return the pot to the hob over a low heat and bring slowly to a simmer. If the meats are brought to a simmer too quickly after being chilled and sliced, the slices can curl around the edges and seize up. Leisurely is the way to do it. The other benefit of heating the meats slowly with plenty of time before dinner is that if they were not as tender as you would like them to be, they will soften more if you give them more time.

As already noted, this dish is often served in two courses: the broth followed by the meats. Of course, if you prefer to have everything in one bowl, there is no law against it. If you decide to do the former, measure 240ml broth per person into a separate saucepan, bring to a boil, and serve in warmed soup bowls over a slice of garlic-rubbed toast. Keep the meats over very low heat while your guests eat the broth course. Serve the meats on a large warmed platter with the *salsa verde* or Cherry Mostarda on the side.

VACUUM PACKED

Vacuum packaging, a method of food preservation in which all of the air is removed from the package before it is sealed, is a popular way to package meats and other foods to extend their shelf life. It has become extremely common to purchase vacuum-sealed meats, even from small, quality purveyors. Vacuum packaging has a number of positive attributes: it aids in retaining moisture in the short term, inhibits spoilage, and limits the chances of cross-contamination. Also, if you are freezing meat, it will help to prevent freezer burn.

But vacuum packaging has drawbacks, as well. It is often more difficult to judge the quality and age of meats that are vacuum sealed. Also, in the process of extracting air from the bag, liquid is extracted from the meat, which then sits in the package. This liquid, known in the meat business as 'purge', can develop an unpleasant flavour in the meat if stored too long. To avoid these downsides, pass over vacuum-packaged meats that contain a large amount of liquid, and look for tightly sealed bags that clearly display their contents.

When you remove meat from a vacuum-sealed package, it is not uncommon for the meat to give off a slight odour. This is not necessarily an indication of spoilage, however. Rinse the meat quickly under cool running water and pat dry. Any unwanted scent should dissipate within a few minutes.

FENNEL-DUSTED PORK SHOULDER STEAKS

Whether you cut your own shoulder steaks (see instructions below) or have your butcher cut them for you, you will find this simple preparation based on a relatively inexpensive cut produces stunning results with a minimum of effort. Serve alongside Fagioli all'Uccelletto (page 327) and a simple rocket salad. **SERVES 4**

2 pork shoulder steaks, about 600g each
1 large clove garlic, lightly crushed
1¹/₂ teaspoons fennel pollen
Fine sea salt

Lightly rub both sides of each steak with the garlic clove, then sprinkle both sides with the fennel pollen and sea salt.

To grill on the barbecue, prepare a medium fire. Alternatively, heat a large cast-iron frying pan over a medium-high heat until hot.

Place the steaks on the grill grate directly over the coals, or on to the hot frying pan, and grill, turning frequently, for about 20 minutes, until a thermometer inserted into the thickest part of a steak registers 65°C. Serve at once.

BUTCHERING PORK SHOULDER STEAKS

Pork shoulder steaks are an excellent cut for the grill that utilise the whole shoulder. After removing the foot and foreshanks from the shoulder (see steps 1 and 2 on page 112), leave the picnic attached to the pork butt. Remove the bottom of the rib cage from the ribs by working the tip of a boning knife between the cartilage of the rib cage and the ends of all four ribs. Stand the whole shoulder up with the square end on your board, the arm bone pointing up, and the ribs facing to the right. About 2.5cm to the left of where the ribs begin, make a cut 2.5cm deep from one end of the shoulder to the other, directly in front of the skin. Lay down your knife and use your hands to push the rib side of the incision gently down towards the table while holding the skin side of the shoulder in place. The shoulder should open up slightly along a natural seam where the cut is. If it seems a little stuck, use the tip of the knife to make the original cut deeper, cutting at a slight angle towards the ribs. Tug again at the shoulder to open it up, then look inside the cut on the skin side to locate the shoulder blade. Starting at the top of the shoulder, carve directly against the shoulder blade while pulling the rib section down towards the cutting board. Once the knife has gone along the inside of the bone all the way to the bottom, turn the knife towards the skin side and cut all the way through the skin, separating the halves. Set the portion with the shoulder blade aside. On the other half, cut between each rib all the way to the backbone. Nestle a cleaver into the first cut you just made, lay a folded towel over the back of the cleaver and tap on the cleaver with a rubber mallet until the blade goes through the bone. (See photos for step 1 on page 103.) Repeat with all of the other ribs. Cut the rest of this piece of shoulder into steaks the same thickness as the rib steaks. The remaining half of the shoulder (which contains the shoulder blade) can be turned into two or three skin-on roasts (simply remove the shoulder blade bone). Alternatively, remove the skin and use this piece for ground pork, stew meat, or sausage.

The Leg

A bone-in whole leg is a phenomenally gorgeous cut, but not terribly practical outside of feeding the festive hordes or preparing your own dry-cured ham. But a boneless leg will yield smaller roasts and leaner meat that works well for making salami.

THE HAM

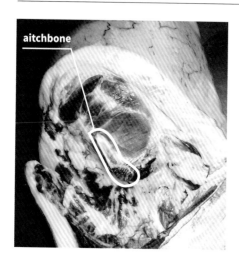

◄ **1. Start to remove the aitchbone.** The trickiest bone to extract from a pig is the aitchbone, which is the bone that attaches to the ball joint on the leg. To remove it, locate the teardrop-shaped bone about 15cm in from where you separated the leg from the middle. Using a boning knife, remove the meat from around the bone where it protrudes above the surface.

▶ **2. Separate the ball joint from the socket of the aitchbone.** Working directly against the aitchbone on the side closest to the trotter, cut roughly 7.5cm into the meat until you hit another bone. This is where the ball joint of the femur bone connects to the socket of the aitchbone. Work the tip of your knife between the ball joint and the socket to separate the two.

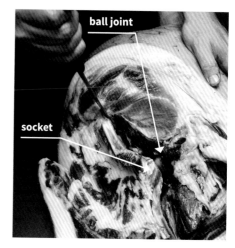

3. Detach the aitchbone. Working simultaneously, use one hand to peel the aitchbone away from the ball joint and the other hand to cut away the leg muscle with a boning knife. (Sometimes, part of the backbone will still be attached to the aitchbone, in which case you should simply cut away the backbone along with the aitchbone.)

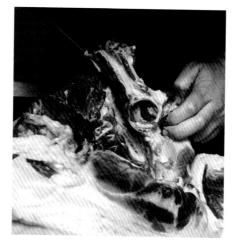

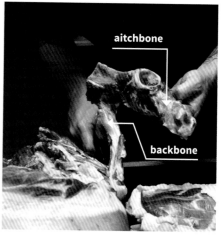

4. Cut off the trotter and hock leg. Saw the back foot off directly below the hock leg, (cut 1, right). Next, measure halfway between the cut you just made and where the hock leg widens somewhat and saw through the bone to make one hock leg cut (cut 2, right). Using a chef's knife, and working parallel to the cut you made to remove the trotter, slice through the skin and meat down to the bone. Use your saw to cut through the bone, then resume cutting the meat and skin with your knife to separate it completely. Make the same cut to yield a second hock where the shank widens considerably (cut 3, right).

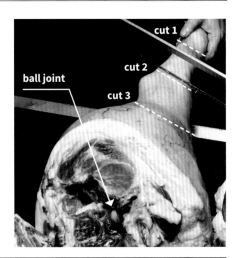

5. Remove the femur bone. Return to the place where you severed the ball joint from the aitchbone. Following the bone, and using a boning knife, cut along one side of the femur bone that attaches to the ball joint until it ends at the patella, the round bone about 2.5cm in diameter that sits next to the intersection of the femur and shank bones. Cut along the other side of the femur, then carve underneath it while pulling it away from the muscle with your free hand.

6. Remove the remainder of the hock leg, the patella, and any remaining bone or glands. Returning to the place where you severed the hock, use a boning knife to carve away the bone. Any hock meat that remains attached to the leg can be cut for sausage or stewing.

7. Cut the leg into smaller roasts. Following the natural seams in the meat, cut the boneless leg into four smaller, evenly shaped roasts. Trim off any excess from each roast and save any scraps for sausage.

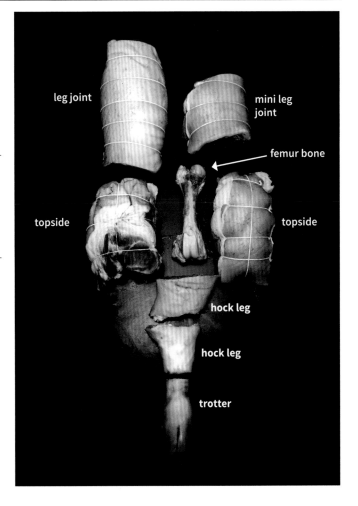

TONNO DI MAIALE

Lean pork leg is transformed into a moist and tender delicacy in this unique preparation with its origins in the Chianti region of Tuscany. *Tonno* is the Italian word for 'tuna'. Similar in texture to oil-packed high-quality tuna, the pork leg in this recipe is salted for a couple of days, then simmered in an aromatic mixture of white wine and water and finally immersed in good olive oil. The result is a brightly flavoured meat with a wonderfully flaky texture, perfect for crostini with roasted cherry tomatoes or for adding to a salad of creamy white beans with red onion and parsley.

Fresh bay is an essential seasoning in this recipe. If you have difficulty locating fresh Mediterranean bay leaves, you can substitute good-quality dried bay leaves, increasing the number of leaves by roughly one-third. **MAKES 4 X 500ML JARS**

3.5kg fine sea salt
1 skin-on, boneless pork leg roast, about 1.5kg
500ml dry white wine
8 fresh bay leaves
About 1 litre olive oil

Select a glass or plastic container just large enough to accommodate the roast. Pour a layer of salt 5cm thick into the bottom of the container. Nestle the pork in the salt. Pour in more salt to surround and cover the pork completely. No part of the roast should be visible. Cover and refrigerate for 48 hours.

Remove the pork from the salt and rinse lightly under cold running water to remove any salt crystals clinging to the meat. Put the pork, skin side up, into a tall, narrow pan. Add the wine and half the bay leaves. Pour in enough cold water to cover the meat. Place the pan over a medium-high heat and bring to a rapid simmer. Skim any impurities that may rise to the surface as you would a broth. Partially cover the pan to allow some steam to escape but to keep the liquid from evaporating too quickly. Cook for about 1½–2 hours, until the skin starts to peel away from the meat. Take the pan off the hob and remove the lid to let the pork cool to room temperature in its cooking liquid.

When the pork has cooled, using a slotted spoon, transfer it to a tray or large plate. Discard the cooking liquid and bay leaves. Use your hands to remove the skin and external fat and discard. They should come off easily. Tear the meat into roughly 5cm chunks.

Thoroughly clean and dry 4 glass 500ml jars with lids. Pour 60ml of olive oil into the bottom of each jar, then press one of the remaining bay leaves against the side of each jar. Pack the pork chunks fairly tightly into the jars, leaving 2.5cm of headspace. Pour in more oil to immerse the meat completely. Poke the meat gently with the tip of a knife to release any air pockets. The oil tends to settle and create space. Wait for 10 to 15 minutes, then pour in more oil if necessary. Seal the jars and refrigerate for at least 2 days before using.

To serve, using a clean spoon or fork, gently remove the pork you plan to use from the jar. If any meat remaining in the jar is protruding above the oil, pour in more oil as needed to cover. Properly covered with oil and refrigerated, the pork keeps well for 4–6 weeks.

Offal and Extremities

Beyond the everyday chops and roast, the pig is endowed with a wealth of edible offal and extremities from its meaty head to its springy tail. These parts make up a large portion of the pig and learning to use them is a noble skill. The charcuterie offers a variety of ways to incorporate offal into sausages and terrines, but they are often delicious enough to warrant a place of honour on a dinner menu, too.

The Head

Nothing says pork-o-phile like the purchase of a pig's head. What some might find gruesome, the rest of us know is good stuff. The head can be used in its entirety to produce an aromatic broth, or its jowls, cheeks, tongue, and ears can be used separately.

Jowls are an irregularly ovular cut off the head. When they are cut into, they reveal a streaky makeup, like a smaller version of pork belly. Sometimes mistakenly labelled cheeks, the jowls actually hang below the pig's jawline. They are sold with or without their skin and can be cured for Guanciale (page 296) or smoked for what is known as face bacon.

Cheeks are the small, dense muscles that rest above the jaw. They are flavourful but tough and should be braised to tenderise them. Use cheeks to make a *ragù* for serving over pasta or polenta, slice them for tacos, or add their meat to terrines.

Tongue is also a muscle. Pork tongues, unlike much larger beef tongues, do not require peeling before eating. They do need a long, slow simmer to render them tender, however. Once cooked, they can be diced and added to terrines, sliced and served as part of an antipasto, or pickled.

Ears are quite cartilaginous. They need to be braised for several hours to become tender. After braising, ears are usually crisped or fried so that their exterior is crunchy and their interior is yielding.

Pig Skin

Pig skin is a marvel. It can be fried into golden sheets of crunchy *chicharrones*, used to coat a roast au naturel with its crispy crackling, or boiled and added to sausages and cooked salamis to add a bouncy texture.

Scoring the Skin: To achieve crispy crackling on the skin of a pork roast, you must score it first. Scoring the skin facilitates the rendering of fat during roasting and allows the release of steam that would otherwise be trapped beneath the skin's surface. If you don't score, this steam produces rubbery, chewy skin. When scoring the skin, it is crucial to use a light touch. Too deep a gash can result in the loss of moisture and flavour. Too shallow and you risk the already-mentioned rubbery finish. Turn a boning knife upside down so the blade faces up and carefully scratch the surface with the tip of your knife. If you see the white of the fat, you are scoring too deeply. As the roast cooks, the skin will pull away from the score marks.

Skinning the Pig: To trim the skin from any cut of pork, use this simple method: Place the pork, skin side up, on a cutting board. Choose any corner of the skin to begin, and slide your knife between the skin and the meat parallel to the board. Cut outwards towards the closest edge of the pork, creating a loose flap of skin. Turn the knife back towards the meat and insert it into the open area between the loose flap of skin and the meat. Now cut along the edge of the roast while holding on to the loose skin, cutting at a slight upward angle to peel back the skin while leaving as much fat behind as possible. When cutting, keep your hand behind the knife, so if the knife slips through the skin, it won't go into your palm.

Trotters and Tails

Lip-smacking trotters and tails are loaded with collagen. A long, slow simmer will break down the tough connective tissues and tenderise the meat. They can be braised with greens for a classic southern treat. An excellent source of gelatine, trotters and tails can also be added to broths for increased body or used to fortify Head Cheese (page 248) and other potted meats. Trotters and tails can be cured or smoked and used to flavour other dishes, as well.

Offal

The heart, liver, and kidneys of a pig are typically stronger tasting than those of their ruminant counterparts. A plate of pork kidneys is not for the uninitiated. However, pork offal plays a crucial role in many charcuterie preparations, where their flavour lends a subtle earthiness to the finished dish.

Offal is highly perishable, so freshly harvested offal is always best. Source your pork offal from a good butcher, preferably one who specialises in whole hogs. Because offal, especially livers and kidneys, is responsible for filtering impurities from the body, it is wise to purchase it from a butcher who handles only naturally raised pork. If you are purchasing a whole hog and would like its offal, be sure to request it specifically, as some slaughterhouses automatically discard it.

Liver is particularly prized for the richness it adds to pâtés and liverwurst. Fresh liver should be eaten within a few days. Liver is often frozen immediately at the time of harvest. If it has been frozen, soaking it in milk overnight will help draw out some of the metallic flavours that can sometimes develop as a result of freezing. When preparing pork liver, be sure to cut away any discoloured spots and connective tissue with a sharp paring knife. Diced liver can be added as a garnish to country-style terrines or can comprise a portion of the forcemeat to add richness. It can also be added to the forcemeat of sausages and salami.

Heart is more of a muscle than an organ meat. It can be used in terrine recipes in place of lean meat. Simply slice it open lengthwise and trim away any excessive blood spots and silver skin. It adds a richness and complexity not found in most other muscles.

Kidneys often have white connective tissue around where they were cut from the pig. Simply trim this tissue away, and if desired, soak the kidneys in milk to tame their intense flavour (this will make them suitable for use in charcuterie). Fresh kidneys from naturally raised hogs are best. Pork kidneys can be diced and used as a garnish for pork terrines, or substituted for liver when liver isn't available.

PIG HEAD POZOLE

Yes, this recipe calls for a pig's head, quartered so that it fits more easily into your pot. Ask your butcher nicely and he or she will cut it for you. Why a whole pig's head? The head has the perfect ratio of sticky skin, bone, cartilaginous bits and face meat to give this traditional Mexican pork stew its delicious, slightly feral quality. There is no equivalent substitution.

A pot of *pozole* makes perfect party fare, and this recipe easily feeds a small crowd. Serve with plenty of cold beer and shots of fiery mescal. If you can't find hominy (dried corn kernels), substitute fresh corn. It won't be authentic, but it will still taste good. **SERVES 12–14**

1kg dried hominy or fresh sweetcorn
2–2.3kg boneless pork shoulder
Fine sea salt
1 skin-on pig's head, quartered
2 dried bay leaves
1 sprig oregano
900g diced yellow onion
6 dried chillies de árbol
6 ancho chillies
6 guajillo chillies
1 large head garlic, separated into cloves

GARNISHES
120g thinly sliced radish
400g shredded green cabbage
170g minced white onion
4 limes, cut into wedges
4 ripe avocados, halved, pitted, peeled, and sliced
12g dried oregano

The day before you plan to cook, cover the hominy, if using, with cold water. Cut the pork shoulder into 4cm cubes, season with salt, coating evenly, place in a bowl and cover. Refrigerate overnight.

The next day place the pig's head, bay, and oregano in a large pan with enough cold water to cover. Place over a medium heat and bring to a simmer. Reduce the heat to a lazy bubble and cook for about 3 hours, until the meat easily pulls away from the bones. Using tongs, transfer the large pieces of head to a plate and let cool. Strain the broth through a fine-mesh sieve into another large pan and add the cubed pork and onion. Place the pan over a low heat, bring to a simmer and cook, uncovered, for 2 hours, until the pork is fork-tender.

While the pork is cooking, pull the meat off of the cooled head pieces and chop into bite-sized pieces. Cut the ears away from the head and julienne. Add the chopped meat to the simmering pan.

Meanwhile, cook the hominy. Drain off the soaking liquid, transfer to a large pan, cover generously with cold water and place over a medium-high heat. Simmer vigorously, uncovered, for 2–3 hours, until the kernels 'flower' and become tender. (If using fresh sweetcorn, boil for 3–6 minutes, until the corn rises to the surface of the pan.) Drain, reserving the cooking liquid.

While the pork and hominy are simmering, toast the chillies in a frying pan over a medium heat for 1 minute, until fragrant and pliable. Turn out on to a plate to cool. Seed the chillies and tear them into smaller pieces. Place in a bowl, add warm water to cover, and soak for about 30 minutes, until softened. Drain, reserving the soaking liquid.

Using the same pan, roast the garlic cloves over a medium heat for about 5 minutes, until the skins darken and the fragrant aroma is released. Turn out on to a plate to cool, then peel.

In a blender, combine the garlic, chillies and 250ml of the chilli-soaking water. Purée until smooth. If the mixture is too thick, add more of the soaking water. Set aside.

Add the cooked hominy or corn to the simmering pot of broth and meat along with enough of the cooking liquid to achieve a good consistency. You want about equal parts flavourful broth to meat and corn. Stir in half the chilli paste and taste for seasoning. Adjust the seasoning with additional chilli paste and salt if needed.

To serve, ladle the *pozole* into soup bowls and leave your guests to garnish.

PICKLED PORK TONGUES

Eating is believing, and you must actually eat the pig's tongue to comprehend how good it can be. The tongue is a well-exercised muscle that can be quite tough. Brining and simmering it before pickling gently softens the muscle and infuses it with tons of flavour. Thinly slice these pickled tongues and use in salads or serve with Horseradish Salsa Verde (page 323) as an hors d'oeuvre.

MAKES 3 PICKLED TONGUES

3 pork tongues

BRINE
4 litres water
225g fine sea salt
115g sugar
1 tablespoon peppercorns
1 tablespoon yellow mustard seeds
3 allspice berries
2 dried bay leaves
10 cloves garlic, crushed

COOKING LIQUID
2 litres pork broth (see Basic Rich Broth, page 44)
120ml dry white wine
2 tablespoons fine sea salt

PICKLING LIQUID
500ml cider vinegar
1 tablespoon sugar
1/2 teaspoon fine sea salt
2 allspice berries
1 teaspoon peppercorns
3 cloves garlic, peeled but left whole
1 onion, thinly sliced

To make the brine, in a large pan, combine the water, salt, and sugar, place over a medium-high heat, and bring to the boil, stirring occasionally to dissolve the salt and sugar. Meanwhile, put the peppercorns, mustard seeds, allspice, and bay on a square of cheesecloth, bring the corners together, and tie securely with kitchen twine (or use a small muslin bag). When the liquid is boiling, add the spice sachet, remove from the heat, and let cool to room temperature.

Rinse the tongues and place them in a non-reactive plastic or glass container large enough to hold the tongues and the brine. Pour in the cooled brine and refrigerate for 24 hours.

To make the cooking liquid, put the broth, wine, and salt into a pan, place over a medium-low heat, and bring to a low simmer. Remove the tongues from the brine, rinse briefly under cold running water, and add to the pan. Cook the tongues gently for 2½–3 hours,

until they are tender when pierced with the tip of a knife. Make sure they are immersed in liquid throughout the cooking process, adding water as needed to completely submerge the tongues.

Remove from the heat, let the tongues cool in the cooking liquid, then refrigerate in the liquid until well chilled. Remove the chilled tongues from the cooking liquid, then pack into a clean 1-litre jar.

To make the pickling liquid, combine the vinegar, sugar, salt, allspice, peppercorns, garlic, and onion in a small saucepan over a medium heat, and bring to a simmer. Cook for 10 minutes to release the flavours from the onion, garlic, and spices.

Remove from the heat and let cool to room temperature. Pour over the tongues to immerse them completely, then cover and refrigerate for 2 days to allow the tongues to macerate fully before serving. The tongues are best if eaten within 2 weeks.

Ruminants: Lamb, Goat, and Beef

Lamb, goat, and beef are all ruminants, a group of tasty four-legged mammals with an inborn preference for a vegetarian diet. There are over 150 ruminant species, but domestic cattle, sheep, and goats account for about 95 per cent of the total population. Ruminants are distinguished from other herbivores by their smartly designed multi-chambered stomach, which includes the all-important rumen that allows them to thrive on grasses, twigs, and shrubs.

Humans have long relied on the ruminant as a source of meat, milk, and clothing. Around 10,000 BCE, we had the grand idea of keeping a few of these easily managed beasts close at hand for our convenience. Sheep, goats, and cattle foraged for their own food, cared for their own young, required little in the way of fencing and, if you decided to pack up your tent and head to the next village, your band of merry ruminants could easily tag along, providing a food source on the way. Over time, as the human population grew, our appetite for these tasty beasts, with their irresistible steaks, succulent chops, and tender loins grew right along with it. We developed a red-meat habit and we were willing to do some less-than-wholesome things to feed it.

The dawn of industrial agriculture saw the boom of the feedlot, or animal feeding operation (AFO). Ruminants require a lot of land on which to forage to reach their ideal weight. Human demand for abundant red meat at low prices was at an all-time high around the middle of the twentieth century. The questionable solution was to speed up production of meat by feeding animals grain near the end of their life cycle. Sheep, goats, and cattle love grain. To them, it is like candy. And like candy, it fattens them, making their meat well marbled. But an all-candy diet is extremely unhealthy. Their little rumens cannot handle large amounts of grain for long. A diet high in grain dramatically lowers the pH in the rumen, which can cause a variety of ailments that must be cured by antibiotics. The confined quarters in which the animals are kept does not help this situation, and the amount of waste produced in high concentrations is enough to make you lose your appetite completely.

Today a greater awareness of the perils of the feedlot system is slipping into mainstream consciousness and the market is providing more choice. Healthy animals produce healthful meat. A good butcher will be able to provide you with grass-fed animals or animals that have been finished for just a short time on more natural grains, such as barley, as opposed to cheaper, less digestible corn. A good butcher will have meat that is hormone and antibiotic free and certified humane. Using a local source for our lamb, goat, and beef helps to reduce the environmental impact of our red-meat habit. Branching out from steaks and chops and learning to use all of the delicious parts of the animal in a thoughtful way can also help to satisfy our red-meat fix and lessen the effects of it on the environment.

Lamb and Goat

Sheep and goats are both members of the subfamily Caprinae, making them practically cousins. In many respects they are quite similar. Many of their differences are apparent in their outward appearance (most sheep are woolly and goats have hair), their eating habits (pickier sheep prefer short, tender grasses and goats are happy dining on just about any shrubs, grass, leaves, or vines), and their social behaviour (sheep love to be part of a flock and the naturally curious goat has an independent streak). But by the time these kin arrive at the butcher's or on your dinner plate, you might be hard-pressed to distinguish between them. The breeds, diet, and lifestyle seem to have more effect on the flavour and fat content of the meat than the species do.

Both have meat that comes in delightful ruby hues. Both whole carcasses usually weigh in around 23kg. From a butcher's perspective, both yield nearly identical cuts. The primary difference is that goat tends to be leaner, but not drastically so. In the charcuterie, the two can be used interchangeably for a variety of sausages and terrines, although some additional fat may be needed for leaner goats. The flavour of the two, especially in more mature animals, is so similar that in some Caribbean and Asian countries the word *mutton* is used colloquially to describe both sheep and goat.

Sheep are descended from the wild mouflon of Europe and the Near East and are one of the oldest species to be domesticated. Today some one thousand distinct breeds exist, a greater number than any other livestock. Sheep breeds are classified by their primary purpose: meat, milk, or wool. Meat sheep are distinguished by their age: a lamb is an animal less than 1 year old, a hogget is a male or female 1–2 years old, and a mutton is a female or a castrated male more than 2 years old. In general, the meat from younger lambs is milder and more tender than the meat of more mature animals. But the meat from well-raised older lambs and from muttons and hoggets that is properly hung or aged can have a richly nuanced flavour, comparable to beef, and still be quite tender.

Goat is the world's most popular source of animal protein, with strong ties to the cuisines of Africa, Asia, the Middle East, the Caribbean, and South and Central America. Today's goat is the descendant of the wild goat of southwest Asia and eastern Europe. It is a lean, nutritious source of meat, with a lower carbon footprint than lamb, beef, or pork, and goat is ever so slowly becoming a more popular fixture on menus and in markets in North America and northern Europe.

Both young goats and the meat of young goats are called kid, and despite its reputation for tasting gamey, the meat has a flavour that is often milder than lamb.

Lamb and Goat at the Butcher's

A quality butcher's should be able to provide good local lamb, although in colder climates, lamb is sometimes raised seasonally. Because goat is less popular in many communities, you will probably have to order it in specially. Avoid purchasing imported lamb and goat, which is nearly always frozen for its long journey across the sea. The meat of both animals can range from rosy to nearly wine coloured but should be vibrant. The fat of both animals should be fairly white, and it will be a bit more abundant on the lamb.

When searching for a good source for lamb, look for a butcher who buys whole lambs. They will be able to offer you a variety of interesting cuts beyond just pricey, popular rack and T-bones. Often overlooked, shoulder is excellent for braises, stews, and brochettes and can also be ground for burgers. Slowly simmered shanks and necks are a cold-weather favourite. Boneless leg is great for the grill. Braised tongues can be thinly sliced and added to a salad.

Lamb and Goat in the Charcuterie

The intense flavour and firm fat of lamb and goat do not lend themselves as easily to charcuterie as other meats, but in cultures where lamb and goat are staples, and especially in cultures where pork is taboo, the tradition of lamb and goat charcuterie thrives. The fattier shoulder, shank, and breast are ideal for sausages and terrines, and the breast can also be cured and smoked in a manner similar to pork bacon. Braised and sliced lamb tongues make a sophisticated terrine garnish.

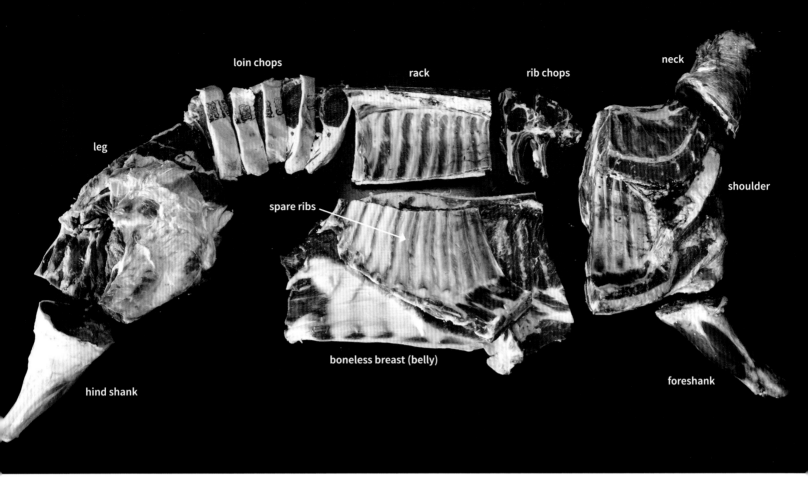

loin chops

rack

rib chops

neck

leg

shoulder

spare ribs

boneless breast (belly)

foreshank

hind shank

Butchering Whole Lamb and Goat

Lamb and goat are straightforward beasts to butcher. They are smaller and more simply put together than a pig. If you are a novice in the world of whole beast butchery, these two animals provide an excellent jumping-off point for you to try your hand at breaking down a whole animal.

The Neck, Shoulder, and Foreshank

Lamb and goat are generally sold headless, so the first three cuts at the front end are the neck, followed by the shoulder, and then the foreshank below that. All three of these cuts are fairly lean and great for stewing and braising. For step-by-step instructions, see pages 134–6.

The Legs and Hind Shanks

Tender yet extremely flavourful legs are great for roasting and grilling. They can be left on the bone, boned, butterflied, or broken down into single muscle roasts or cubed for brochettes. The hind shanks, like the foreshanks, are best suited to braising and stewing. See pages 137–9 for instructions.

The Middle

The middle consists of the primal loin, rib rack, and breast or belly. Left whole, the primal loin and rib rack are perfect for roasting or grilling. Alternatively, they can be cut into chops and pan seared or grilled. The belly or breast is perfect for use in the charcuterie, where it makes a great addition to lamb sausage or terrines, or can be cured for lamb bacon. For step-by-step instructions, see pages 140–1.

1. Cut off the neck. The neck bends where it connects to the shoulder. Using a boning knife, make a cut down to the bone. Use a saw to cut through the bone, then finish cutting with the boning knife to sever the neck completely. The neck can be left whole or cut into cross sections for braising.

2. Remove the shoulders. Lay the goat on its back. Using a boning knife, cut between the fourth and fifth ribs on one side of the goat, starting at the rib cage end. (If you have difficulty cutting through the bottom of the rib cage, use a saw to cut through the bones, then resume using the knife.) Make a long, smooth cut between the fourth and fifth ribs, all the way from the rib cage down to the backbone.

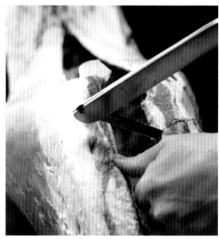

3. Repeat on the opposite side. Cut between the fourth and fifth ribs on the other side of the goat, all the way from the rib cage to the backbone (again, using the saw to cut through the rib cage if necessary). Once you get to the backbone, use the saw to cut through it completely. The shoulders should now be completely severed from the middle.

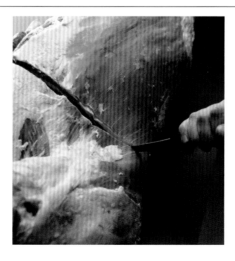

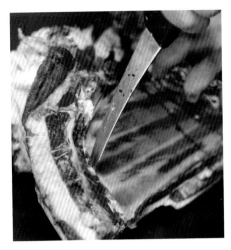

4. Cut off the foreshanks. The arm bone that connects the shank and the shoulder forms a right angle at the joint. Make a cut with your boning knife at the joint and saw through the bone to remove.

5. Saw apart the shoulders. Separate the shoulders from each other by sawing directly through the backbone.

6. Bone the shoulder (optional). Separate the ends of the rib from the breastbone.

7. Score each side of the ribs all the way down to the backbone, then run your knife behind each to free it completely from the muscle. As each bone is exposed, twist it until it pops out of its socket in the backbone and cut it away with the knife.

8. Remove the rest of the backbone once all of the ribs have been removed.

CONTINUED

9. Next, remove the shoulder blade. Find the end of the shoulder blade, located roughly in the centre of the shoulder. Holding the knife parallel to the cutting board, slip the knife between the meat and the bone. Run the knife along the top of the blade, directly against the bone. As you cut, peel the meat from the top of the shoulder blade towards you. The shoulder blade will begin to narrow until it meets the arm bone. Locate the socket that connects the two with your finger. Using the tip of the boning knife, free the shoulder blade from the ball joint of the arm bone, then begin to outline the sides of the blade with your knife back towards the widest part of the blade.

10. Cut against the bone on the underside of the shoulder blade. Roughly halfway through, the bone forms a slight downward ridge. Follow the ridge with the knife. Cut close to the bone along each side of the ridge until it separates from the muscle completely.

11. Remove the arm bone where it was connected to the shoulder blade by cutting down each side of it with the knife. Then cut the underside, leaving behind as much meat as possible. Using your boning knife cut away the remaining piece of backbone.

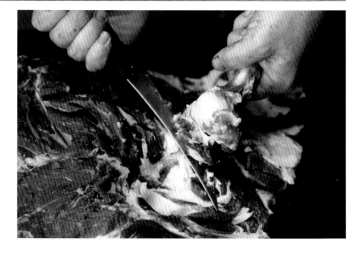

1. Remove the tenderloins. The tenderloins are the two long cylindrical muscles that sit on the inside of the backbone in the abdominal cavity. If the lamb or goat has its kidneys intact, cut them away with a boning knife to make the tenderloins more visible.

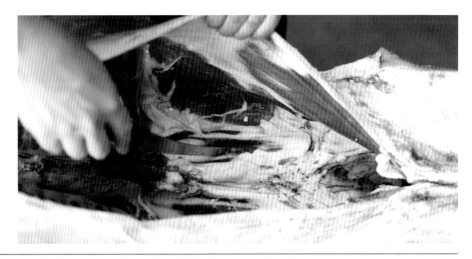

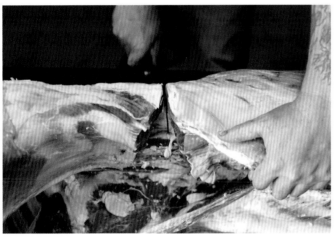 

2. Now separate the legs from the middle section. Look at the lamb from its side and note where the top of the leg curves. Roughly three-quarters of the way down the length of the tenderloins from the shoulder end, cut away from the rib cage through the flank. Saw through the backbone. Repeat the initial cut on the other side to remove both legs.

▶ **3. Saw apart the legs.** Separate the legs from each other by sawing directly through the backbone.

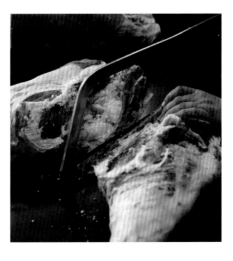

CONTINUED

4. Remove the aitchbone. To remove the aitchbone from a leg, locate the teardrop-shaped bone 10 to 15cm in from where you separated the leg from the middle. On the side closest to you, using the boning knife, remove the meat from around the bone where it protrudes above the surface. Working directly against the bone on the side closest to the shank, cut roughly 5cm into the meat until you hit another bone. This is where the ball joint of the femur bone connects to the socket of the aitchbone. Work the tip of the knife between the ball joint and the socket to separate the two. Then, using your free hand, peel the aitchbone away from the ball joint while cutting away the leg muscle with the knife. For a bone-in leg roast, trim any excess fat from the edges of the leg, particularly around the flank. The leg can be trussed if desired.

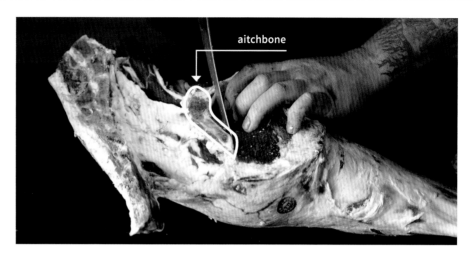

aitchbone

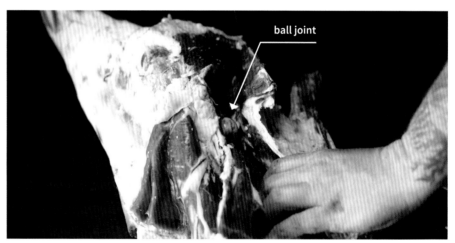

ball joint

socket of aitchbone

5. Remove the hind shank. If you prefer boneless roasts, begin by removing the hind shank where it bends at the knee. Bend the knee back and forth a little, then, using your fingertip, find the gap between the femur and the shank bone. Separate the hind shank with your knife, making as straight a cut as possible. If you have difficulty making the cut with the knife, saw through the bone, being careful to make a clean cut.

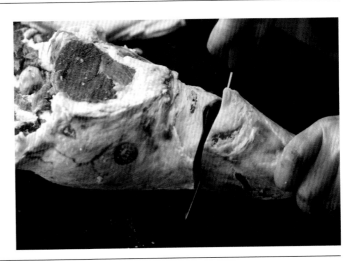

6. Bone the legs (optional). Remove the femur from the leg by cutting directly along the bone down one side and back up the other. Scrape the underside of the bone to separate it as cleanly as possible. Separate the leg into individual roasts by cutting at the natural seams between the muscles, making sure to remove any large glands or blood vessels as you go. The glands that must be taken out are yellowish to dark grey and will most often reside in thick pockets of fat between the muscles. The individual roasts can be trussed for uniformity. Be sure to save any trim for lamb sausage.

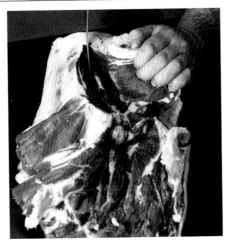
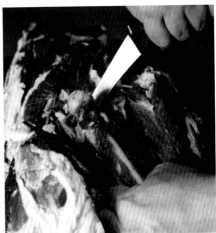

1. Split the middle. Saw lengthwise through the backbone of the middle section to separate the left and right halves.

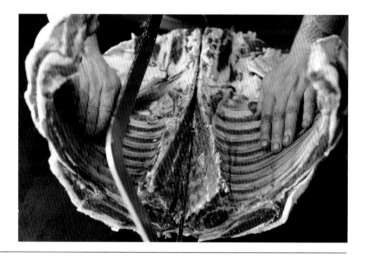

2. Separate the breast from the primal loin. Saw through the rib bones parallel to the backbone, roughly 20cm towards the bottom of the ribs. Once you have sawed through all of the ribs, use the boning knife to make one long cut from one end of the middle to the other, following the separation of the ribs.

3. Bone the breast. Begin at the shoulder end where the rib sits flush against the edge of the breast or belly. Using the boning knife, and cutting parallel to the work surface, carve the ribs away from the belly, leaving a little meat on them if you plan on braising or smoking them. Save the breast meat for braising or for sausage or terrines.

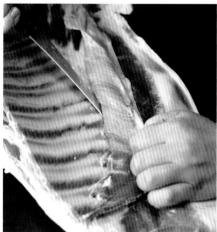

4. Separate the loin from the rib rack. Using the boning knife, cut directly behind the last rib down to the backbone. Using a heavy cleaver with a towel folded over the top and a rubber mallet, cut through the backbone to separate the loin from the rib rack: holding the handle of the cleaver in one hand, use your other hand to tap the back of the cleaver with the mallet until the blade goes through the bone completely.

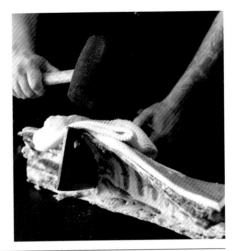

5. Cut the short loin into chops. Using the cleaver and mallet, cut along the backbone at roughly 5cm intervals to break the loin into four or five chops.

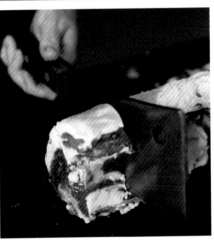

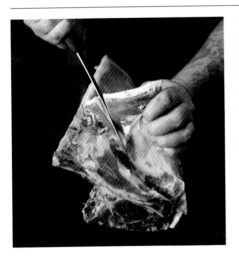

◄ **6. Remove the shoulder blade from the rib rack.** Locate the remaining bit of shoulder blade that sits behind the first three ribs. Using the boning knife, cut away the shoulder blade, leaving as much meat as possible behind.

► **7. Cut rib chops.** Cut between each rib to portion into individual chops using the cleaver and mallet method described in step 4.

LAMB RIB CHOPS WITH RAS EL HANOUT

Ras el hanout is a classic North African spice blend. Loosely translated it means 'head of the shop', symbolising the best the spice merchant has to offer. Traditionally, the blend contains a panorama of different spices, usually about twenty but sometimes many more, and each merchant has his or her own recipe. Although we generally advocate toasting and grinding your own spices, we love the perfectly balanced blend available from our neighbours at Whole Spice in Napa. In their version, allspice, bay leaf, black pepper, white pepper, cardamom, cinnamon, chilli, clove, cumin, ginger, mace, nutmeg, saffron, rosebuds, and more are carefully combined and ground to produce a truly aromatic mix perfect for seasoning lamb. **SERVES 4**

8 lamb rib chops
Fine sea salt
2 cloves garlic, crushed
2 dried bay leaves
1 tablespoon thinly sliced orange rind
2 tablespoons extra-virgin olive oil
1 tablespoon ras el hanout

Season the lamb chops on both sides with salt.

In a large, shallow bowl, combine the garlic, bay, orange rind, and olive oil and mix well. Add the lamb chops to the bowl and turn as needed to coat evenly. Sprinkle the *ras el hanout* over the lamb chops and again turn as needed to coat evenly.

To grill the chops on a barbecue, prepare a medium-hot fire. Place the chops on the grill grate directly over the coals. Grill, turning once, for about 2 minutes on each side for rare or about 3 minutes on each side for medium-rare to medium. Serve at once.

Alternatively, to cook on the hob, heat a large cast-iron frying pan over a medium-high heat until hot. Add the chops and cook, turning often, for about 5–7 minutes total. Serve at once.

TO FRENCH OR NOT TO FRENCH?

For presentation, many butchers remove a good portion of the backbone (or chine) from the rack and 'french', or trim all of the meat and fat from the top portion of the ribs to expose the bone. Although this gives the rack a dramatic look, it lessens its succulence. We prefer to keep the chine attached and leave a little meat and fat clinging to the rib bones. Meat cooked next to the bone is always more delicious. Removing the chine on the inside of the loin exposes the delicate muscle to dry heat as it roasts, which increases the chances that it will go from perfect to overcooked within seconds. When you leave the chine attached, it provides you with a bit of insurance. The chine can easily be carved away after cooking, before you slice and serve the rack. The meat along the rib bones is also a treat. Peeling the excess away robs you of the delights of gnawing all that rib meat away once the chop meat has been devoured.

GOAT SHOULDER BIRRIA

Whenever we get a whole goat shoulder, we fix a pot of *birria*, ice down some beers, and invite over a few friends. *Birria*, which translates literally as 'a mess', is a spicy Mexican stew traditionally prepared for holidays and special occasions. The marriage of toasted chillies, heady spices, and flavoursome goat meat is a happy one, well worth celebrating. Good fresh goat can sometimes be tricky to procure, so be sure to arrange with your local butcher ahead of time. Lamb shoulder or lamb shanks can be used in place of goat. Prepare the meat one day ahead. **SERVE 6-8**

1 bone-in goat shoulder, 2.3–2.7kg

Fine sea salt

4 ancho chillies

4 guajillo chillies

6 dried cascabel chillies

1 teaspoon black peppercorns, toasted
    (see page 11)

1 teaspoon cumin seeds, toasted
    (see page 11)

2 whole cloves

2.5cm piece cinnamon stick

2 serrano chillies, seeded and chopped

150g chopped onion

4 cloves garlic, coarsely chopped

1 teaspoon dried Mexican oregano

60ml fruity vinegar such as cider or plantain

1 x 360ml bottle lager

1kg plum tomatoes, roasted, seeded, and
    puréed

TO SERVE

60–85g minced white onion

8–10g chopped fresh coriander

Plenty of tortillas, warmed

Lime wedges, for serving

One day ahead, season the goat shoulder. Rub the meat liberally with salt and set the shoulder in a wide bowl.

Toast the chillies in a cast-iron frying pan over a medium heat for about 1 minute, until fragrant and pliable. Turn out on to a plate to cool. Seed the chillies and tear them into smaller pieces. Place in a bowl, add warm water to cover, and let soak for about 30 minutes, until softened. Drain, reserving the chillies and soaking liquid separately.

Using the same pan, toast the peppercorns, cumin, cloves, and cinnamon over a medium heat for about 5 minutes, until fragrant. Turn out on to a plate to cool, then transfer to a mortar or spice grinder and grind finely.

In a blender, combine the softened chillies, serranos, onion, garlic, oregano, ground spices, and vinegar and purée until smooth. If the mixture is too thick, add a few tablespoons of the reserved chilli soaking liquid to thin to a good paste consistency. Coat the goat shoulder with the chilli paste, then cover and refrigerate overnight.

The following day, position a rack in the lower third of the oven and preheat to 135°C (gas 1).

To cook the stew, you will need a large, sturdy pan or casserole fitted with a steamer insert or a rack. Pour the beer into the pan. Set the goat shoulder on the rack or in the steamer basket. Scrape any extra chilli paste over the meat.

Set the pan over a medium heat and bring the beer to a simmer. Cover and place in the oven to steam slowly for about 4 hours, monitoring the progress about once every hour. If the beer begins to evaporate, add 240ml boiling water to the pan. Remove from the oven when the meat pulls easily away from the bone.

Transfer the meat to a platter and remove the steamer insert. Ideally, you will have about 720ml cooking liquid left in the bottom of the pan. If not, add enough water (or broth) to make up the difference. Add the tomatoes to the pan, place over a low heat, and simmer for about 20 minutes, until slightly thickened. Taste for seasoning.

Slip the bone out of the shoulder and cut or shred the meat into large chunks.

To serve the *birria*, place portions of the goat meat in deep bowls and ladle the sauce over the meat. Top each bowl with 1 tablespoon each of the minced white onion and coriander. Serve with the tortillas and lime wedges. Alternatively, serve the birria at the table with tortillas and garnishes on the side.

Beef

Beef, source of endless cravings and deep hankerings, is the delicious, blood-red meat of domestic cattle. Cattle are the descendants of the now-extinct aurochs that once grazed the grasslands of Europe, North Africa, and Asia. The ancestors of today's modern herd were most likely domesticated somewhere in the Fertile Crescent about ten thousand years ago, to provide ancient humans with milk, leather, draft power, fertiliser, and, of course, tasty, tasty beef.

Cattle are grazers, consuming vast amounts of grass and requiring more land than any other domestic animal. And unlike modern pork and poultry farming, for the majority of their lives, cattle are still raised outdoors on smaller family farms, often referred to as cow-calf operations. In many cases, when their calves are weaned and of a reasonable size, small-scale farmers sell them off to larger concerns or feedlots, where they are 'finished' on a diet of grain, mainly corn, that helps to pack on the pounds.

In some cases, however, the cattle are allowed to mature on pasture for an additional 6–12 months, sometimes with a small supplement of more natural grains. That's the good beef. Cattle that have been allowed to mature slowly will naturally produce beef with depth of flavour and with characteristics influenced by their breed and their diet.

Beef in the Butcher's

A good butcher will source good beef. If you want the best quality, look for beef that has lived its life naturally on pasture without hormones and antibiotics. Look for a shop that uses whole carcasses, quarters, or at least primal cuts of beef. It will have a superior selection of cuts, will have the ability to custom cut, and will most likely be grinding fresh beef daily. If you are lucky, the beef may even be aged in-house.

Beef Cuts

Cattle are big creatures, and the world of beef cuts is vast. There are often two, three, or even four different ways to butcher the same primal cut, which can make things even more perplexing. We like to think of beef in terms of how we will be preparing it. Do you feel like eating a plate of carpaccio, a steak off the grill, or a warming bowl of beef stew? Often several cuts can serve a similar purpose. The final preparation should be your guide in purchasing.

Raw Beef

Freshly cut raw beef that is thinly sliced or finely chopped is a delicious treat with culinary roots around the world. Always seek out the best-quality beef, especially for raw preparations, and always make sure it is perfectly fresh. In general, beef for raw preparations should be fairly lean and relatively tender. Be sure to tell your butcher you plan on eating it raw so he or she can find you the best cuts.

Tenderloin: The name says it all. This tender, oblong-shaped muscle spans two primal cuts: the short loin and the sirloin. For carpaccio, steak tartare, and other raw beef dishes, select a cut from the large end, closest to the sirloin. Be sure to trim away any silver skin from the outside, which is unpleasantly chewy.

Sirloin: Lean, relatively tender, and flavourful, sirloin is a good alternative to pricier tenderloin and works well in all raw-beef preparations.

Eye of Round: This lean but well exercised and often tough cut is from the centre of the silverside. Although it is tougher than the loin and sirloin, it can be thinly sliced and lightly pounded to tenderise it enough to serve it raw. Or, it can be finely chopped and served raw. This flavourful cut works well in such dishes as Vietnamese *bo tai chanh*, a raw-beef salad in which the beef is first marinated in citrus juice, or Ethiopian *kitfo*, a dish of chopped raw beef tossed in a warm spice butter.

Beef Seared, Roasted, or Grilled

Leaner cuts, cuts from less-used muscles, and thin cuts do best seared, roasted, or grilled using hot, dry heat to produce meat with an irresistible brown exterior, preserving a tender, rosy interior.

Rib Roast: This celebrated beef roast, a festive favourite, is comprised of ribs six to twelve in its entirety, but can be cut into smaller (two- to six-rib) roasts. This massive cut should always be seasoned several days in advance of cooking and brought to room temperature before cooking. The rib roast can be cooked on the grill or in the oven and is usually started at a very high heat and then allowed to finish at a lower temperature. If you are serving standing rib carved off the bone, be sure to brown the bones a little longer before serving.

Rib Eye or Entrecôte: This well-marbled steak is from the rib section and is comprised of the longissimus muscle and the buttery spinalis, or cap. A grilled rib eye is the steak that cowboys dream of.

T-Bone: A steak cut from the striploin, the T-bone is made up of the fillet and sirloin, connected with a T-shaped bone, similar to a porterhouse but with a smaller section of fillet.

Porterhouse: Like the T-bone, the porterhouse steak includes the fillet and sirloin connected with a T-shaped bone, but with a greater portion of tenderloin. This glamorous cut often gets top billing on steak house menus. But buyer beware: the bone running through the centre can make the porterhouse difficult to cook evenly.

Top sirloin: More marbled than its neighbouring tenderloin, this cut, relatively low in connective tissue, is quite tender. Many prefer this cut to the fattier rib eye. Steaks cut from the back end, closer to the rump, tend to have more connective tissue and can be tougher.

Tenderloin: This very lean, long muscle tucked beneath the ribs and next to the backbone is well suited to fast, high-heat cooking. Whole tenderloins are available unpeeled (fat and silver skin intact), peeled (fat has been removed but silver skin remains), or as PSMO – peeled, side muscle on. The tenderloin can be further dissected into the fillet, a cut from the narrow end of the tenderloin nearest the rib, and the chateaubriand, a thick cut from the centre or large end of the tenderloin closest to the sirloin. If you plan on cooking a tenderloin whole, tuck the thinner end under itself and tie it as you would a roast to ensure that it cooks evenly. Lean tenderloin is often paired with a rich sauce such as béarnaise.

Rump: Lean, flavourful and tender, rump is found behind the striploin and can be cut into boneless steaks or roasts. If you are roasting a whole rump, be sure to remove the interior sinewy seam before seasoning and tying.

Tri-Tip: A small, triangular-shaped boneless cut from the bottom of the sirloin, the tri-tip likes a lengthier span on the grill, fat side down, over a lazy fire.

Eye of Round: This typically quite lean and moderately chewy cut from the centre of the silverside works well if roasted hot and left fairly rare. Because it is particularly thick and dense, it should be seasoned a day or two in advance.

Bavette: A classic bistro steak cut close to the flank, this relatively thin cut requires a hot, fast sear or grill. It can be used as a substitute for skirt steak.

Thin Flank: A long, flat, lean cut from the abdominal area behind the brisket, flank needs to be cooked quickly at high heat to rare or medium-rare to avoid becoming ropey.

Skirt: A long, flat, flavourful cut from the inside of the rib cage, skirt steak calls for very high heat in order to brown without overcooking. It makes excellent *carne asada* tacos.

Onglet: This long, tubular cut comes from the interior plate section that hangs from the diaphragm near the liver. Onglet has a rich, almost livery flavour and is a bistro favourite, perfect for *steak frites*. When slicing cooked onglet, note that the grain changes direction as the meat tapers.

Thick Rib (Thin Cut): Cut from the rib and plate, thin-cut thick ribs are cut so that they need only a hot, fast grill to cook them through. A strongly flavoured marinade works best with these ribs.

Flat Iron: Cut from under the shoulder blade, the flat iron is flavourful and cooks very evenly.

Topside: Not the most marbled or tender roasting cut, topside makes a reliable roast beef.

Beef Braised, Stewed, or Smoked

Cuts of beef that are well exercised, tough, and full of connective tissue are rendered tender and delicious when cooked slow and low using either a moist cooking method such as stewing or braising, or a low-heat smoke. Always be sure to salt these cuts at least a day ahead (longer for larger cuts) to ensure the seasoning fully penetrates the meat.

Neck: Beef neck, great for stewing, is generally cross cut, similar to oxtail. A long, slow simmer makes the neck meat tender and succulent.

Chuck: This large shoulder cut yields an abundance of stew meat and pot roasts. Marbling can vary on chuck, so look closely for well-marbled pieces. The eye of the chuck, the circular muscle closest to the loin, is preferred for pot roasts.

Brisket: Cut from the breast area and consisting of the animal's pectoral muscles, brisket is loaded with connective tissue and, if properly smoked or braised, can be extremely tender and delicious. Brisket should have a thick fat cap on one side that helps to keep the meat moist during cooking. Cook the brisket with the fat cap side up to protect the meat.

Thick Ribs: This cut, which comes from both the plate and rib, is rich, meaty, and tender when browned and braised. The whole thick rib is generally cut one of two ways, either English cut (into short, thick lengths) or flanken or long cut (across the bones).

Leg: This leg cut is tough and sinewy and needs a long, slow simmer to render it tender. Leg can be purchased on or off the bone. Boneless leg can be cut into stew meat.

Oxtail: Literally the animal's tail, this bony, gelatine-laden section is usually cross cut, perfect for braising or stewing.

Silverside: Lean and tough silverside does not succumb easily to tenderness. Thinly sliced, it can be rolled and braised as in *braciole* or smoked and dried for jerky.

The Price of Dry Ageing

Dry aging beef for a period after slaughter makes for tastier, more tender meat. It allows for some loss of moisture, which concentrates flavour, and provides the naturally occurring enzymes time to break down the tough connective tissues. Better beef is usually hung for 1–2 weeks. Certain cuts, such as the rib and short loin, benefit from an even longer period of about 4 weeks. This allows time for the development of fungi that create the distinctive dry-aged flavour.

For a butcher, dry ageing beef is costly in terms of labour, materials, and real estate. The beef will lose a good amount of weight due to evaporation and because the exposed surface areas must be trimmed. That means that the butcher must sell it for a significantly higher price per pound than meat that has not been aged the same way. But if you love the taste of good dry-aged beef, it is well worth the price tag; just follow the 'eat less quantity but higher quality' ethos.

Beef in the Charcuterie

Beef doesn't figure as prominently in the charcuterie as pork or duck. Still, chuck, thick rib, and brisket, all flavourful cuts from the forequarter with evenly dispersed fat, are excellent for use in sausage and in meat loaves (page 254). Brisket can also be brined and smoked for pastrami (page 304). The eye of round is often cured and air-dried for *bresaola* (page 300). Gelatine-laden oxtails can make a rich potted meat (page 250). Because beef fat is very firm, beef must typically be ground twice when used for sausage and salami, such as our Pepperoni (page 238).

Butchering Beef

Although cattle are essentially larger versions of goats and sheep, muscularly speaking, they are roughly ten times the size, and beef butchery is a more intricate process. Good beef butchery is an art form. If you are considering trying your hand at beef butchery, a class or internship with a professional is a good place to begin. Or, you can start small and practise butchering subprimals and primals.

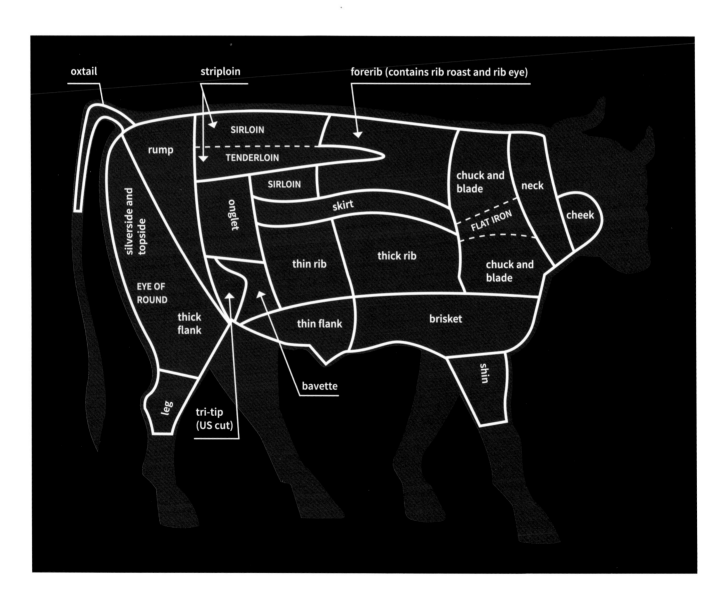

PEPOSO

This is one of the most delicious, simple, and low-maintenance stews ever devised. Its origins lie in the medieval Tuscan town of Impruneta, just south of Florence, famous for its terracotta tiles. The tile-makers would fill a pot with cheap cuts of beef, season the meat with pepper and garlic, add wine to cover, and then slide the pot into the still-warm kiln when they finished firing for the day. This version includes the modern addition of tomato purée, but the idea is the same. The meat does not require browning, and you can use any beef trimmings you have on hand as long as they are somewhat consistent in size. You can slip the pot into the oven before you hit the sack, and you will wake the next morning to a perfectly cooked pot of tender beef stew.

Traditionally, this stew is served warm with nothing more than good bread and more wine, but it is also quite good spooned over pasta, polenta, or mashed potatoes. **SERVES 6**

1.5kg boneless stewing beef (such as chuck, leg, or brisket), cut into roughly 4cm cubes

12 cloves garlic, peeled but left whole

2 tablespoons tomato purée

1 tablespoon fine sea salt

3/4 teaspoon chilli flakes

3/4 teaspoon freshly ground pepper

750ml bottle red wine (such as Chianti Classico or Sangiovese)

Preheat the oven to 110°C (gas ¼).

Place the beef, garlic, and tomato purée in a casserole dish or Dutch oven. Sprinkle the salt, chilli flakes, and pepper evenly over the beef and garlic and stir to coat evenly. Pour the wine into the pot.

Place the pot on the hob over a medium heat and bring to a lively simmer. Cover the pot and transfer to the oven. Cook for 8 hours, until the meat is meltingly tender.

Depending on the cuts used, the stew may need to be defatted. This is done most easily when the stew is chilled. Uncover the pot and allow the stew to cool to room temperature. Then refrigerate it for 2–3 hours to allow the fat to solidify. When the fat has hardened on the surface, lift it off and reserve it for another use, then reheat the stew to serve.

BARBECUE-GRILLED RIB EYE FOR TWO

What better way to express your love than to cook a fat, sexy rib eye over a glowing wood fire? A thick, dry-aged rib eye, also known as the cowboy steak in the US, is a cut that says you are willing to take the time to do things right and willing to pay the price for both love and good meat.

You will need to start preparing your rib eye two days in advance, but plan even further ahead. Begin with a phonecall to your butcher to inquire about procuring the most beautiful, dry-aged, bone-in rib eye he or she can cut for you. Let them know it is for love. All butchers are romantics at heart.

Accompany the beef with good bread, a simple salad, and a worthy bottle of wine. **SERVES 2**

1 bone-in whole rib eye steak, preferably dry aged, about 7.5cm thick
Fine sea salt
2 tablespoons olive oil
Fleur de sel, for serving

Season the rib eye 2 days in advance. Sprinkle the salt from on high, so that it distributes evenly over the steak. Be sure to get the sides as well as the fat on the outer rim. Cover loosely with a clean kitchen towel (it needs to breathe), and refrigerate overnight. The following day, turn the steak over, re-cover and refrigerate for 1 more day.

Remove the rib eye from the fridge 2 hours before grilling. The closer it is to room temperature, the more evenly it will cook and be medium-rare throughout. Massage 1 tablespoon of the olive oil into each side of the rib eye.

Prepare a hot barbecue. Oak, mesquite, or almond wood burn well and provide a delicate perfume that enhances the flavour of the beef. Scrub the grate if it has been a while. No soot shall soil your soon-to-be-perfect steak.

When you have hot coals (you should be able to hold your hand 15cm above the fire for no more than 3 seconds), lay the steak on the grate. After 5 minutes, lift the steak gently, rotate it 90 degrees, and set it back down on the grill. After another 5 minutes, turn it over and repeat.

Lastly, turn the steak to rest on its side on the grill to crisp the edge. Insert a thermometer into the thickest part of as close as possible to the bone without touching it. For rare, remove the steak at 40°C, for medium-rare, 43°C. For anything above 43°C, this is the wrong recipe and the wrong steak.

Transfer the steak to a cutting board, cut away the rib and put it back on the grill to brown and crisp. Let the steak rest for 8–10 minutes before carving. Wet the table with your good steak knives and a dish of *fleur de sel*.

Using a sharp knife, cut the steak into three sections following its natural seams. Start by removing the teardrop-shaped muscle at the bottom (the cap, or spinalis dorsi), then the triangular-shaped piece of fatty meat at the top of the steak. This leaves the large circular piece in the middle known as the eye. Slice each piece crosswise through the muscle so that you have two identical pieces of each muscle. Fetch the browned rib off the grill.

To serve, place 1 piece of each of the muscles on each plate. You will have to share the rib – or fight over it.

CARNE CRUDA

After a long transcontinental flight with missed connections and a jarring car ride from Nice, we finally arrived in adorable Alba too late for lunch. We found one tiny restaurant about to shut its doors for the afternoon that took pity on us. The kitchen was officially closed, but the staff fixed us a plate of *carne cruda*, a slightly intimidating heap of hand-cut raw beef drizzled with olive oil and accompanied with a lemon wedge. It was a love-at-first-bite moment and, to our surprise, we polished it off with gusto, then proceeded to eat our weight in various incarnations of *carne cruda* throughout Piedmont.

Both well-trimmed sirloin and tenderloin, two cuts that are lean, flavourful, and tender, work well in this recipe. It is crucial to use fresh high-quality beef and to cut the meat by hand. To ensure a small, uniform dice, chill the beef thoroughly beforehand and use a sturdy, sharp chef's knife. Crisp flatbreads, crostini, or halved hard-boiled eggs make excellent accompaniments. **SERVES 6 AS AN HORS D'OEUVRE OR 4 AS A FIRST COURSE**

500g lean beef sirloin or tenderloin, well chilled

1 teaspoon coarse sea salt (such as Maldon or fleur de sel)

$1/4$ teaspoon finely ground pepper

1 clove garlic, finely chopped

Juice of 1 lemon

1 tablespoon finely chopped fresh flat-leaf parsley

Generous handful of rocket leaves, cut into narrow ribbons

75ml extra-virgin olive oil, plus more for drizzling

Trim any silver skin, gristle, or large pieces of fat from the exterior of the beef, then cut into 0.5cm cubes. Place the beef in a bowl. Add the salt, pepper, garlic, lemon juice, parsley, and half of the rocket and fold them into the beef to mix evenly. Stir in the olive oil and taste for seasoning.

To serve, mound the beef on to a large plate. Garnish with the remaining rocket and drizzle with olive oil.

RARE ROAST BEEF

In the charcuterie, the eye of round, a long, lean, tough cut of beef tucked away in the silverside, is usually turned into *bresaola* (page 300), a cured, air-dried cut of beef. Eye of round is also delicious roasted rare. The beef is seasoned 2 full days ahead of roasting to allow the flavours to penetrate this thick cut, and the result is a tender, rosy roast. When purchasing eye of round for roasting, ask your butcher for an untrimmed one. The fat will help to keep the meat moist while cooking. This beef is not cooked to temperature – when it's brown, it's done. If you know your oven runs hot, keep an eye on the beef so it doesn't overcook.

We like to serve warm slices of this rare roast drizzled with good balsamic and topped with a lightly dressed rocket salad. It can also be roasted ahead and served chilled, accompanied by horseradish sauce or aioli. Leftovers make great sandwiches. **SERVES 6–8, WITH LEFTOVERS (DEPENDING ON APPETITES)**

2 cloves garlic, peeled but left whole

Fine sea salt

1 untrimmed eye of round, about 2.3kg

1 juniper berry

1 tablespoon peppercorns

2 dried bay leaves

1 tablespoon dried sage

1 tablespoon dried rosemary

1 tablespoon brown sugar

60ml olive oil

Using a pestle and mortar, pound the garlic cloves with 1/2 teaspoon salt until crushed to a paste. Rub the crushed garlic all over the beef, then salt the beef generously with additional sea salt, roughly 3 tablespoons for a 2.3kg roast.

In a spice grinder, combine the juniper berry, peppercorns, bay, sage, and rosemary and grind finely. Transfer to a small bowl, add the sugar, and stir to mix well, then rub the mixture over the beef. Wrap the beef tightly in cling film and refrigerate for 48 hours.

Take the beef out of the fridge and leave it for about 2 hours to come to room temperature before roasting. Preheat the oven to 220°C (gas 7).

Place the meat on the rack of a roasting tin. Rub the roast with olive oil and place in the oven. Roast for 10 minutes, then rotate the pan 180 degrees and roast for an additional 8 minutes. Remove from the oven and let rest for 10–15 minutes before slicing.

4

SKEWERED, ROLLED,
TIED & STUFFED

When you yearn for more than just a plain roast or chop, pick up your knife, grab your ball of butcher's twine, and learn to manoeuvre beyond the basics of butchery. At the Fatted Calf, we pride ourselves on making succulent skewers, stuffed birds and beasts, and seasoned roasts that range from the classic to the unusual. These specialities are as beautiful to behold as they are to savour at the table. Mastering a handful of nifty skills will allow you to butcher, season, cook, and carve with flourish. Those techniques, combined with marinades and seasonal stuffings, will help you to elevate simple hunks of meat to impressive centrepieces for all occasions.

Skewered

Cooking meat on skewers is almost as old as lighting a fire. It is hard to resist the ancient allure of meat on a stick. Quick and versatile, it is perfect for feeding as few as two or as many as twenty.

Most cuts of meat, from duck breast to goat legs, can be skewered. Although lean and tender cuts are often favoured, even those cuts not traditionally associated with high-heat grilling, such as shoulder cuts of pork, beef, goat, or lamb, can work equally well. They are flavourful, especially when marinated ahead of time, and if you have not lost the ability to chew your food, they are a good choice for skewering. Avoid using overly tough cuts with an abundance of connective tissue, such as shank and neck cuts. Cut or cube the meat into similar sized pieces, usually 2.5–4cm, to ensure uniform cooking. You want plenty of surface area for browning (which improves flavour), but the pieces must not be so small that you lose the juicy, succulent centre. Season the meat before skewering to allow the flavours to develop. For leaner meats, a few hours will suffice. For more robust cuts (such as lamb or pork shoulder), a day or two is best.

A variety of options exist for spearing any meat, from bamboo sticks to lengths of sugar cane. Although most are merely a vehicle for the meat, some, such as lemongrass or sturdy herb stems, are meant to impart flavour. Reusable stainless steel skewers are preferred for most jobs. Because they are good heat conductors, they speed the cooking, and unlike wooden skewers, they never splinter or burn. That said, bamboo skewers definitely have their place, especially at parties or around the campfire. To prevent the exposed wood on bamboo or other wooden skewers from burning, soak them in cold water for at least half an hour before threading meat on to them, and be sure to position a piece of meat over the tip of each skewer.

Half an hour before you are ready to skewer, remove the seasoned meat from the fridge to come to room temperature. This will help it to cook more evenly. Thread the pieces of meat through their centre or thickest part on to the skewer, spacing them at least 2.5cm apart. Although it may look more appealing to have the meat pieces tightly packed on the skewer, spacing them generously helps the meat to brown more thoroughly and cook more evenly. Likewise, decorating your meaty skewers with a cornucopia of seasonal vegetables makes them pretty and colourful, but meats and vegetables seldom cook at the same rate and should be skewered separately. If you are cooking different cuts on the same skewer, be sure to choose ones that will cook at more or less the same rate. More delicate cuts, such as lean poultry, rabbit, or organ meats, benefit from a protective layer of fat wrapped around each piece. Caul fat or thinly sliced bacon or pancetta works well. Thread oddly shaped or unwieldy pieces of meat on to two parallel skewers to prevent the meat from flopping around during cooking.

Although you can cook skewers under a grill or in a grill pan, there is no real substitute for cooking over a fire outdoors. Cook skewered meats over relatively high heat, around 290°C, or roughly the same temperature you would use for a steak or chop. If you are grilling over charcoal, place your skewers on the grill the moment the flames die down, leaving active coals that radiate a large amount of heat. Turn your skewers every minute or so to brown on all sides. Skewers cook quickly, most of them taking no more than 8–10 minutes.

PORK BROCHETTES WITH HERBES DE PROVENCE

The alluring scent of these pork morsels, crusted in black pepper, white pepper, and *herbes de Provence*, is hard to resist as it wafts off the grill. These brochettes, a summertime favourite at the Fatted Calf, are an excellent choice for a large gathering. The recipe can easily be doubled or tripled, and the meat can be seasoned well in advance. **SERVES 6**

1.2kg boneless pork picnic, cut into 4cm cubes
2¹/₂ teaspoons fine sea salt
4 cloves garlic, lightly crushed
1 onion, sliced 1cm thick
2 tablespoons olive oil
¹/₂ teaspoon black peppercorns
¹/₄ teaspoon white peppercorns
2 tablespoons Herbes de Provence (page 15)

Season the pork with the salt. In a large bowl, combine the garlic, onion, and olive oil. Add the seasoned pork and turn to coat evenly.

Grind the peppercorns and herbs finely in a spice grinder. Dust the mixture over the pork, then mix to coat evenly. Cover and refrigerate for at least 1 day and preferably for 2–3 days. If you are marinating for longer than 1 day, be sure to mix the pork once a day to redistribute the seasonings.

Prepare a medium-hot barbecue. Thread the pork cubes on to skewers, spacing them about 2.5cm apart. Discard the onions, garlic, and other remnants of the marinade.

Grill the skewers, turning frequently, for 8–10 minutes, until evenly browned on all sides.

Alternatively, cook the skewers under a preheated grill, turning them frequently, using the same timings.

HARISSA-MARINATED LAMB KEBABS

Harissa is a North African hot sauce used to spice up a variety of dishes, meaty and otherwise. Although it is usually served as a condiment alongside food, it also makes a wonderfully spicy marinade for fuller-flavoured meats such as goat, beef, or lamb. If you are a spice fiend, mix up a little extra *harissa* to drizzle over your skewers at the table. Serve the kebabs over couscous or wrapped in flatbread, accompanied with grilled vegetables and lemon-scented yoghurt sauce.
SERVES 6

1.2kg boneless lamb leg, cut into 4cm cubes

2¹/₂ teaspoons fine sea salt

HARISSA

1 teaspoon cumin seeds, toasted and
   ground (see page 11)

1 teaspoon freshly ground black pepper

1¹/₂ teaspoons crushed dried Aleppo pepper
   flakes (pul biber)

1¹/₂ teaspoons unsmoked Spanish paprika

¹/₂ teaspoon ground cayenne pepper

¹/₂ onion, sliced 6mm thick

3 cloves garlic, crushed

1 tablespoon olive oil

In a large bowl, season the lamb with the salt and mix well.

To make the *harissa*, in a small bowl combine the cumin, black pepper, Aleppo pepper (*pul biber*), paprika, cayenne, onion, garlic, and olive oil, and stir to incorporate. Pour the *harissa* over the seasoned lamb and mix well. Cover tightly and refrigerate for at least 1 day and preferably for 2–3 days. If you are marinating for longer than 1 day, be sure to mix the lamb once a day to redistribute the seasonings.

Prepare a medium-hot barbecue. Thread the lamb cubes on to skewers, spacing them about 2.5cm apart. Discard the onion, garlic, and other remnants of the marinade. Grill the skewers, turning frequently, for 6–8 minutes, until evenly browned on all sides.

Alternatively, cook the skewers under a preheated grill, turning them frequently, using the same timings.

MARSHA'S GRILLED RABBIT SPIEDINI WITH OLIVES, ALMONDS AND LEAF SALAD

This recipe was handed down to us from Marsha McBride, chef and proprietress of Berkeley's Café Rouge. Before 'sustainability' became a household word and nose-to-tail eating became fashionable, Marsha was sourcing quality meats from local farms for her shop and using old-world techniques to produce a wide range of charcuterie for her restaurant. Many Bay Area butchers and charcutiers (Taylor included) worked a spell at Café Rouge, where one of our favourite menu offerings is this savoury grilled rabbit *spiedini* (Italian for 'skewers') that calls for both the meat and the organs of the rabbit, served over a leaf salad with olives, and almonds. **SERVES 4**

1 rabbit, 1–1.5kg, boned (see page 86) and
    cut into 2.5cm cubes
2 rabbit kidneys, each halved
1 rabbit liver, quartered
Fine sea salt
1 teaspoon coriander seeds, toasted and
    ground (see page 11)
Grated zest of ¹/₂ orange
2 tablespoons chopped fresh flat-leaf
    parsley
2 tablespoons chopped fresh thyme
1 tablespoon chopped garlic
120ml extra-virgin olive oil
120ml rosé wine
4 thin slices pancetta, home-made
    (page 295) or shop-bought

SALAD

1 shallot, finely minced
60ml red wine vinegar
Fine sea salt
1 tablespoon Dijon mustard
180ml extra-virgin olive oil
80g rocket leaves, stemmed
80g trimmed curly endive hearts
80g trimmed frisée leaves
40g chopped green picholine olives
55g chopped toasted almonds
Freshly ground pepper

Season the meat, kidneys, and liver with salt and the coriander. In a large bowl, combine the orange zest, parsley, thyme, garlic, olive oil, and wine and mix well. Add the seasoned rabbit and mix to coat evenly. Cover and refrigerate overnight.

To make the vinaigrette for the salad, in a bowl, macerate the shallot in the vinegar for a few minutes. Add a pinch of salt and the Dijon mustard, then slowly whisk in the olive oil to emulsify. Set aside.

Prepare a medium-hot barbecue. Remove the rabbit from the marinade. Have ready 4 skewers. Wrap the liver pieces with the pancetta. Thread 1 piece of liver and 1 kidney on to each skewer along with a quarter of the meat, spacing the pieces 1cm apart. Grill the skewers, turning frequently, for 8–10 minutes, until evenly browned on all sides.

Alternatively cook under a preheated grill, turning frequently, for 8–10 minutes.

In a large bowl, combine the rocket, curly endive, and frisée, drizzle with the vinaigrette, and toss to coat. Add the olives, almonds, and a few grinds of pepper and toss again.

Divide the salad among 4 plates. Place a whole skewer, still warm, over each salad and serve.

Rolled and Tied

Most large cuts of meat that will be roasted or grilled benefit from being rolled and tied. This creates a uniformly shaped roast that is easier to cook to perfection, carve, and serve. Left au naturel, unevenly shaped cuts or cuts made up of several smaller muscles, such as pork shoulder or lamb leg, will cook unevenly and their extremities will dry out. On a more practical level, tying keeps your seasonings secured when the meat is cooking. And from an aesthetic point of view, a rolled and tied roast has a bit more professional panache. Attention to the details of cutting, seasoning, tying, and cooking your hunk of meat will result in a roast that is as stunning to behold as it is delicious to eat.

Roasting, Slow Roasting, and Pot Roasting

Roasting, simply defined, means cooking meat in hot, dry, indirect heat, generally in an oven, to produce flavourful caramelisation on its surface. You can roast most meats, from chicken through pork tenderloin to a whole beef rib, and what you choose to roast will dictate how you roast it.

In general, the standard roasting process for a whole bird or a large cut of meat (such as a pork rack or lamb leg) is to get it brown, then turn down the heat. Sear the roast in a hot oven, usually preheated to 190°C–220°C (gas 5–7), to brown the outside, then cook it at a gentler temperature, from 150°F–180°F (gas 2–4), to finish. A stint at a lower temperature ensures the roast cooks evenly. For smaller, leaner, or thinner cuts, such as tenderloin, this second phase of low-temperature cooking is unnecessary.

Slow roasting, which is cooking at a lower temperature, such as 140°C–170°C (gas 1–3), for a longer period of time, with or without an initial hot sear, achieves a different effect. A cut with some fat and connective tissue, such as a lamb shoulder or pork butt, can be slow roasted for several hours until it is thoroughly cooked but still juicy, extremely tender, and nearly falling off the bone.

Pot roasting uses a deeper vessel to retain moisture, which makes it similar to braising, though with a single, large cut of meat. You can either simmer the meat in its own juices or add liquid to the pot to facilitate the cooking. Tougher, more sinewy cuts, such as shank and shoulder, become meltingly tender when pot roasted.

Readying Your Roast

Season well and season early! Some people fear that salt, even a restrained amount, will dry out a roast. True, salt does draw moisture from meat, but after a period of time, thanks to reverse osmosis, the cells reabsorb much of that moisture along with the salt and whatever other aromatics have been used to season, resulting in more flavourful meat. Salt also helps to soften the proteins in meat, making it more tender. Large cuts, cuts with a good bit of fat, meat on the bone, and dense or sinewy cuts greatly benefit from seasoning a day or two ahead. More tender, boneless, smaller or thinner cuts, or cuts with very little fat, can still be seasoned ahead but need to sit for only a few hours. The salt in a marinade or dry rub facilitates the transfer of the flavours to the interior of the meat. A smear of rendered fat or olive oil will help seasoning adhere and provide a little protective coating and browning ammunition for leaner cuts.

All Tied Up

Not every roast needs tying. Many are naturally suited to seasoning and popping in an oven, such as pork loin and other bone-in cuts that have their own built-in structural supports. But roasts that have been butterflied and seasoned, are stuffed, or have an irregular shape require a bit of reassembly and securing with butcher's twine. This ensures that your stuffing doesn't tumble out during cooking and gives the roast a more symmetrical shape so it cooks more evenly.

How to Tie a Roast

Start with a good spool of 100 per cent cotton butcher's twine with a minimum 12-ply thickness (although 16-ply is better if you can find it). Thicker twine is less likely to snap when you pull it taut. When you are working with twine, secure it to your work surface to stop it rolling over, off, and on to the floor while you are in the middle of your *coup de grace* butcher's knot. Place the ball or cone of twine either on a heavy-bottomed base equipped with a dowel or inside a small, heavy saucepan.

Lay your seasoned or stuffed roast on the work surface. If you are tying a roast with an irregular shape, tuck in any protruding bits and pat it into a more even shape. Have your twine close at hand.

Leaving one end attached to the spool, slide the twine underneath the roast. For a rib roast, plan to knot the twine between the first and second ribs; for

a boneless roast, you will make your first knot 5cm or so in from the end.

To make the first butcher's knot (also known as a slip knot), bring the loose end of the twine over the top of the roast. Grab the end of the twine that is still attached to the spool with the ring and little fingers of your left hand (the string will drape across your index and middle fingers), then drape the loose end of twine over the index and middle fingers of your left hand (see photo 1, opposite).

Grab the loose end of twine with your right hand. Then, rotate your left hand so that your palm is now facing down (see photo 2). Since you are still holding the loose end of twine with your right hand, this rotating motion should form a loop on the loose end of twine (not the one attached to the spool). Slip the loose end of twine (which is in your right hand) under and through this loop (see photo 3), then move your left hand out of the way so you can tighten the knot (see photo 4). Hold the loose piece of twine steady in your right hand, then tug the end still attached to the spool tightly with your left hand until the knot is as tight as possible (see photo 5). Make a second, regular knot on top of the first one to ensure the first one doesn't loosen (see photo 6), then cut the twine directly above the second knot (see photo 7).

Repeat until the roast is snugly tied. For a rib roast, tie the twine in between each rib. For a boneless roast, repeat the knot at regular 5cm intervals (see photo 8).

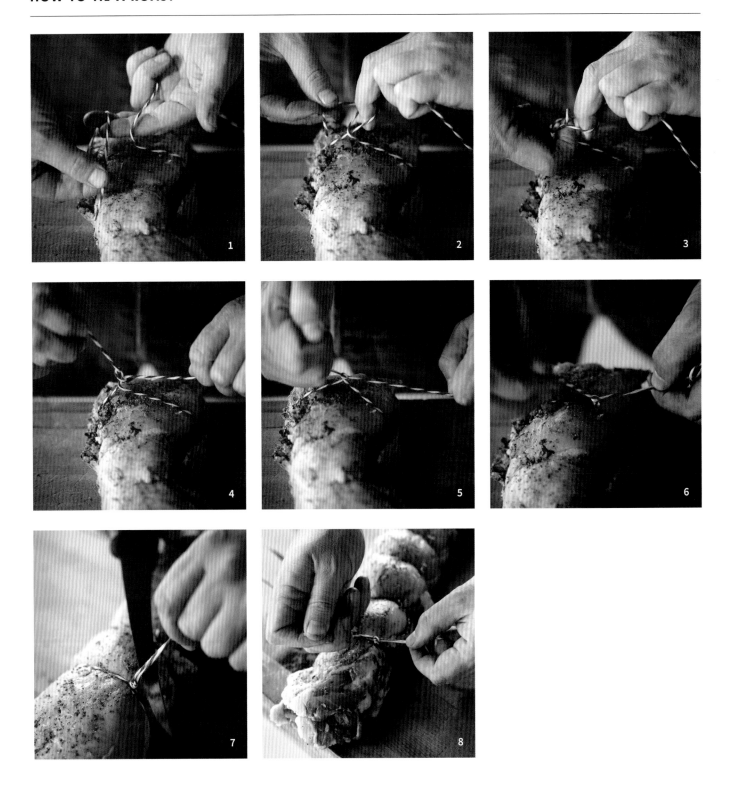

Racks, Pans, and Pots

A solid roasting tin is a good investment. Heavier tins, preferably made from stainless steel or enamelled cast iron, tend to transfer heat more efficiently. Choose a tin that is a bit larger than the roast itself. For smaller, leaner roasts, such as a pork tenderloin or rack of lamb, the tin does not need to be very deep. For a large or fatty cut, such as a pork middle, you want a deeper tin to collect the drippings.

Almost every roasting recipe you read directs you to place the roast on a rack inside the roasting tin. A rack prevents the roast from sticking to the pan, elevates the roast above the pan drippings, and increases airflow for more even roasting. Ideally, the dimensions of the rack almost match those of the tin, mostly for your own safety. If the rack is smaller than the tin, it can slide when you remove the pan from the oven, resulting in hot, sloshing fat that can cause a painful burn. If you do not have a metal rack, you can fashion a rack out of vegetables, such as leeks or carrots, to support the roast during cooking. This vegetable rack can be eaten alongside the roast or blended with the drippings for a robust sauce.

Pot roasts need a good heavy pan. A deep, enamelled cast-iron casserole or Dutch oven transfers heat evenly. Choose a pan that is a bit larger than your roast. If the roast is too snug, it will steam and stew rather than slow roast to a beautiful golden brown.

Time and Temperature

All meat, roasts especially, should be allowed to come to room temperature before cooking. Never transfer a roast directly from the fridge to the oven or you will end up with an overcooked exterior and an undercooked interior. Bringing meat to room temperature reduces the total cooking time and ensures that the meat cooks more evenly. The bigger and denser the roast or cut, the longer it will need. A

CALIBRATE YOUR MEAT THERMOMETER

Even the best meat thermometers can get dropped, banged about in a kitchen drawer, or left to overheat on the hob. It is always a good idea to check to make sure that your meat thermometer is properly calibrated, or reading temperatures accurately. Bring a pan of water to the boil on the hob. Fill a glass with ice cubes and add cold water just to cover the cubes. Place the probe end of the thermometer into the pan of boiling water. It should read 100°C. If not, make a minute adjustment with long-nose (snipe) pliers, carefully turning the nut on the backside of the thermometer, located where the face meets the probe. Then plunge the probe into the ice water. If it does not register 0°C, make another small adjustment until it reads correctly.

We don't recommend using digital instant-read thermometers. They are difficult to calibrate, though some do have a reset button.

chicken may only need half an hour, but a bone-in pork shoulder roast should sit for a couple of hours.

We are frequently asked, how long will it take to cook this roast? The truthful answer is the answer that no one ever wants to hear: we really don't know. We can only guess. When did you remove the roast from the fridge and what temperature is the meat when you put it in the oven? Is your oven perfectly calibrated or does it run hot or cool? Do you have a conventional or a fan oven? Is it gas or electric? Will you be opening the oven door every few minutes or are you more of a set-it-and-forget-it kind of cook? Do you like a little pink in your meat or are you a fan of crispy end bits? Your answers to all these questions will affect the equation.

Two more things will help to ensure a perfect piece of cooked meat almost every time. The first is simple but non-negotiable. Buy a meat thermometer: learn to use it, take care of it, keep it calibrated and you will have no unpleasant surprises – no raw chicken breasts or depressingly overdone legs of lamb. Meat is costly. Don't leave it to chance. We have been cooking meat professionally for almost twenty years and we still use a meat thermometer every day. The second way to improve your meat-cooking skills is to practise. Not everyone is born with razor-sharp culinary instincts. The more you practise roasting (and using your meat thermometer), the more attuned you will be to the look, smell, and feel of a perfectly cooked roast.

Grilling a Roast

When it is too hot to turn on the oven, when the great outdoors beckons, or when there are just too many cooks in the kitchen, take your roast outside and grill it on the barbecue. A roast grilled on the barbecue will take a bit longer, but the exposure

to the irresistible perfume of wood smoke makes the time spent worthwhile.

Use the same principles you would use to cook your roast in a conventional oven. Unlike barbecuing, which calls for cooking at a very low temperature (about 85°C) and with a large amount of smoke to an internal temperature way beyond well-done, a grilled roast should be seared over fairly hot coals then finished slowly, with only a small amount of smoke, until it reaches its target internal temperature. This results in a roast with a balanced, meaty flavour and only a hint of smoke.

Rake the hot coals into a pile slightly larger than the diameter of your roast to create a hot spot for searing. Place your seasoned and tempered roast directly on the grill grate. Place the roast over the hot spot and brown it, turning it frequently, until it is evenly coloured on all sides. Once the meat is evenly and deeply browned, move it off the direct heat and close the barbecue lid. If you don't have a lid, a loose foil tent or large bowl placed over the roast will do. Covering the roast exposes all sides to the trapped heat of the coals so that it cooks slowly and evenly, just as it would in an oven. Turn the roast every now and again until it reaches the desired internal temperature.

Give It a Rest

All roasts, great and small, benefit from a resting period after roasting and before carving. The rare or less-cooked centre will have a few moments to catch up thanks to the help of carry-over cooking. The outer layer of meat has the opportunity to reabsorb some of the meaty juices. The end result is a roast that is more evenly cooked, more tender, and juicier.

PANCETTA-WRAPPED PORK TENDERLOIN

Delicate pork tenderloin is the extra-lean meat adjacent to the loin. This tasty, supple little muscle has a tendency to dry out quickly during cooking and can go from perfectly pink to dry and mealy the second you turn your back on the sauté pan. Our solution is to provide the tenderloin with a protective layer of much-needed fat in the form of rich, salty pancetta. As the tenderloin roasts, the pancetta crisps, basting the lean muscle with its fat so that it remains moist and infusing it with its robust flavour. Roasted radicchio makes an excellent accompaniment. **SERVES 2–3**

1 trimmed pork tenderloin, about 450g

Fine sea salt and freshly ground black pepper

55g Dijon mustard

2 tablespoons white wine

3–4 tablespoons (about 10g) finely chopped fresh rosemary

About 55g thinly sliced pancetta, home-made (page 295) or shop-bought

Roasted radicchio, to serve

Preheat the oven to 220°C (gas 7).

Season the tenderloin on all sides with salt and pepper. In a small bowl, stir together the mustard and wine. Using a pastry brush or your hands, cover the tenderloin liberally with the mix. Sprinkle the rosemary evenly over the roast.

Lay a 25cm-square sheet of waxed or parchment paper on a work surface. Neatly cover the paper with the pancetta slices, overlapping them by about 1cm. Lay the tenderloin 2.5cm in from the edge of the sheet closest to you, placing it parallel to the edge.

Fold the bottom 2.5cm of the paper over the tenderloin, and then roll the paper around the tenderloin. The pancetta should be tightly wrapped around the tenderloin. Remove the paper.

Place the roast on the rack of a roasting tin and roast for about 20 minutes, until the pancetta is golden and crisp and a meat thermometer inserted into the centre of the thickest part of the roast registers 60°C.

Remove the meat from the oven and let rest for 5–10 minutes. Slice into rounds 2.5cm thick and serve with the roasted radicchio.

PORK BLADE RIB ROAST WITH SHERRY, GARLIC, THYME, AND PIMENTÓN

Pimentón is Spanish for paprika, the intensely red spice made from ground dried peppers. The exceptional *pimentón de la Vera*, from the La Vera valley in the Extremadura region of western Spain, is made by hanging the local sweet and spicy red peppers from the rafters of large smoke houses, where they gently dry over fires of holm oak. The resulting ground *pimentón* is classified either as *dulce* (sweet), *agridulce* (medium-hot), or *picante* (hot). All three make a perfect seasoning for pork. Use whichever level of heat you find most satisfying. The country rib (see page 115) works particularly well in this preparation because its natural pocket, directly behind its ribs, allows you to season the meat with the *pimentón* from within. **SERVES 4**

1 pork blade rib roast, about 1kg
2 tablespoons dry sherry
Sea salt
3 tablespoons pimentón de la Vera
8 cloves garlic, sliced paper-thin
2 tablespoons chopped fresh thyme

Butterfly the pork blade roast by making a cut directly behind the ribs from the top to the bottom. Using a pastry brush or the tips of your fingers, brush the roast inside and out with the sherry, then sprinkle salt over its entirety. Lay the roast fat side down on the cutting board and open it like a book. Coat the inside of the roast with the *pimentón*, seasoning from on high so that it disperses more evenly. Next, sprinkle the garlic slices on the inside of the roast opposite the ribs, followed by a shower of the thyme.

Tie the roast with butcher's twine, making knots in between each rib as well as one across the loin (see pages 168–9). Wrap the roast tightly in cling film and refrigerate overnight.

Remove the roast from the fridge and allow it to come to room temperature for about 1 hour. Preheat the oven to 190°C.

Place the roast on the rack of a roasting tin, then place in the oven for 30–40 minutes, until nicely browned. Turn down the heat to 150°C (gas 2) and roast until a thermometer inserted into the thickest part of the roast away from the bone registers 60°C.

Remove from the oven and let rest for 10–15 minutes. Cut away the twine and slice the roast between each rib.

RABBIT PORCHETTA

Porchetta is usually made from a whole hog or whole pork middle, boned, seasoned generously with garlic, fennel, and herbs, then rolled, tied, and roasted. It is a thing of beauty, a great big thing, best cooked for a big, hungry crowd. This miniature version, made with rabbit, follows the traditional *porchetta* principles of rolling a roast with spices, garlic, citrus, and herbs. While still loaded with porchetta flavour, it is scaled for everyday dining and can be prepared in a fraction of the time. **SERVES 6**

1 whole rabbit, 2–3kg, deboned (see
    pages 89–90)
4 cloves garlic, pounded to a paste in
    a mortar
Grated zest of 2 lemons
Sea salt and freshly ground pepper
1 tablespoon fennel pollen
30g chopped mixed fresh herbs (such as
    rosemary, oregano, flat-leaf parsley,
    and sage)
2 tablespoons extra-virgin olive oil

Lay the boneless rabbit out on a work surface and rub the inside with the garlic and lemon zest. Season inside and out with salt, pepper, and the fennel pollen, then sprinkle the herbs over the inside. Working lengthwise, roll up the rabbit tightly, then tie with butcher's twine at 8cm intervals (see pages 168–9). Wrap tightly in cling film and refrigerate for at least 1 day or for up to 3 days.

Remove the roast from the fridge and allow it to come to room temperature for 1 hour. Preheat the oven to 220°C (gas 7).

Place the rabbit on the rack of a roasting tin and rub all sides with olive oil. Roast for about 45 minutes, until a thermometer inserted into the thickest part of the rabbit registers 60°C.

Remove from the oven and let rest for 10 minutes. Slice into rounds 1cm thick. Strain the pan juices and spoon over the sliced meat.

BRASATO AL MIDOLO

In this opulent Tuscan version of beef pot roast, lean, sinewy beef shank is enriched with its own bone marrow, then simmered in delicately sweet, amber-coloured *vin santo*. As the roast simmers, the buttery, beefy marrow oozes throughout, flavouring both the meat and the broth laden with caramelised shallots – an ideal braise for a chilly winter evening! Slice the meat, arrange the slices over mashed root vegetables or polenta, and top with the shallots and marrow-enriched sauce. **SERVES 6**

1 whole beef shank, about 4kg
Fine sea salt and freshly ground pepper
15g fresh rosemary leaves
Sea salt and freshly ground pepper
2 tablespoons extra-virgin olive oil
1kg shallots or cipollini onions, peeled but
    left whole
1.5 litres meat broth, any kind (see Basic
    Rich Broth, page 44)
500ml bottle vin santo

Ask your butcher to bone the shank, cutting along the natural seam between the two main muscles of the shank to yield a single piece of meat. Then ask that he or she split the bone in half lengthwise to expose the marrow.

Season the meat liberally on both sides with salt and pepper. Lay the beef, exterior side down, on a work surface. Sprinkle the rosemary evenly over the inside of the shank. Using a butter knife, pry the marrow from the shank bone (save the bone for your next batch of broth) and arrange it lengthwise along the centre of the roast. Fold the sides of the roast around the marrow to recreate the original shape and secure with several loops of butcher's twine (see pages 168–9). Wrap in cling film and refrigerate overnight to allow the seasonings to penetrate the meat.

Remove the roast from the fridge and allow it to come to room temperature for 1 hour. Preheat the oven to 150°C (gas 2).

In a heavy pan or casserole dish, heat the olive oil over a medium heat. Add the shank roast and brown well on all sides. Remove the roast to a platter and set aside. Add the shallots to the pan and sauté for about 5 minutes, until they are golden. Nestle the shank in the shallots and add the broth and wine. The wine and shallots should reach about three-quarters of the way up the sides of the roast. Bring the liquid to a simmer. Transfer the pot to the oven and cook the meat slowly, turning it occasionally, for 5–6 hours, until it is fork-tender and the liquid is fairly reduced.

You can serve the meat immediately, but like most pot roasts, it tastes best if it is allowed to rest overnight in its braising liquor. Once the pan has cooled to room temperature, cover and refrigerate. The following day, transfer the cold roast to a cutting board. If the fat congealed on the surface of the braising liquid is excessive, remove it. Cut the roast into slices 3cm thick and return the slices to the pan. Heat the pot gently on the hob or in the oven, then taste the liquid for seasoning. Serve the hot slices topped with the marrow-rich sauce and shallots.

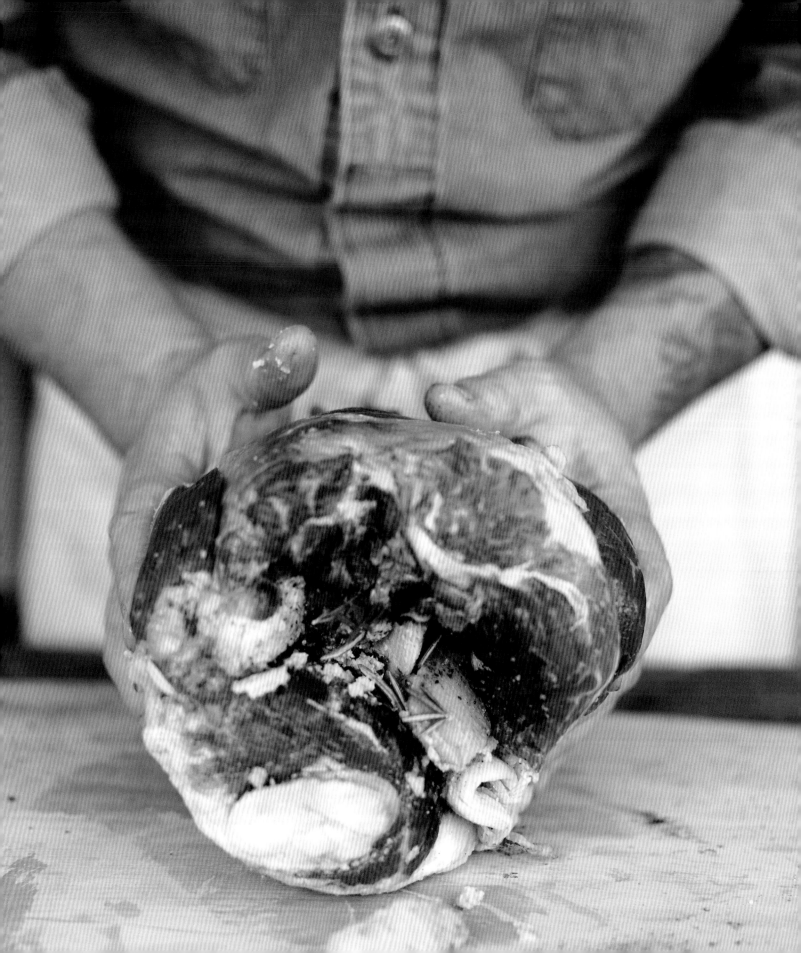

THE CUBAN

Whole, boneless, skin-on pork middle is a cut we generally reserve for *porchetta*, the traditional Italian pork roast rubbed with garlic, showered with fennel and herbs, then rolled, tied, and roasted until its skin turns mahogany and so crunchy it breaks off into terrifically greasy, impossible-to-resist shards. The Cuban is a new-world approach to an old-world classic: same delicious pork middle, same beautiful brown and crispy crackling, but seasoned with a vibrant mixture of cumin, orange, chilli, and herbs. **SERVES 10–12**

½ **skin-on pork middle, 4.5–5.5kg, boned following the instructions for Whole Boneless Middle (pages 106–7)**

170g garlic, pounded to a fine paste in a mortar

Grated zest of 1 orange

Fine sea salt

2 tablespoons black peppercorns, toasted and ground (see page 11)

5 allspice berries, toasted and ground (see page 11)

25g cumin seeds, toasted and ground (see page 11)

55g medium-hot (agridulce) pimentón de la Vera

15g finely chopped fresh oregano

15g finely chopped fresh flat-leaf parsley

2 tablespoons extra-virgin olive oil

75g coarse sea salt

Lay the boned pork middle on a cutting board and rub the garlic into the meat, being sure to get into the grooves where the ribs were removed and into the pocket where the shoulder blade was removed. Spread the orange zest over the meat, then liberally season with fine sea salt followed by the pepper, allspice, cumin, *pimentón*, oregano, and parsley.

Although you can tie the roast alone, it is easier to tie it if you can enlist an extra set of hands. Roll the loin towards the bottom of the belly, pressing it tightly as you go. Tie the roast perpendicular to the light score marks, using double lengths of twine at 5cm intervals (see pages 168–9). Refrigerate the roast uncovered, or loosely covered with a tea towel, for at least 1 day or for up to 3 days to allow the flavours of the seasonings to penetrate the meat. Do not wrap in cling film, as moisture trapped between the plastic and the skin can prevent the skin crisping in the oven.

Remove the roast from the fridge and allow it to come to room temperature for 1–2 hours. Preheat the oven to 190°C (gas 5).

Place the roast on the rack of a roasting tin. Rub it on all sides with the olive oil, then sprinkle it evenly with the coarse salt. Transfer to the oven and roast for 30–40 minutes, until the skin is golden brown and crispy. Turn down the oven temperature to 150°C (gas 2) and continue to cook until a thermometer inserted into the thickest part of the loin registers 60°C. This can take 2½–4 hours, depending on the thickness of the roast.

Remove from the oven and let rest for at least 30 minutes before carving. Cut into slices to serve, using a serrated knife to cut through the crunchy exterior of the skin.

Stuffed

If you want to add a little *ooh là là* to your dinner table, consider the possibility of stuffing your main course. Everybody likes to discover the prize inside, especially when that prize is a hot caramelised fig or a satisfying mouthful of garlicky greens. Stuffings have the added benefit of keeping roasts moist, flavourful and juicy during cooking.

What to Stuff

The traditional Christmas turkey is the best-known stuffed fowl, but numerous other birds, from a plump quail to a flavoursome guineafowl, are just waiting to be stuffed. A whole bird is ideally built for stuffing: the breast cooks much faster than the legs, so stuffing the cavity helps to keep the breast moist while the legs finish cooking. Stuffed boneless or partially boneless birds make an elegant presentation at the table.

Whole beasts such as rabbits, sheep, and hogs are also good candidates for stuffing. Similar to whole birds, their legs and shoulders cook at different rates from their middles. Stuffing their abdominal cavities helps to keep their delicate loins juicy while their legs and shoulders finish roasting.

In truth, you can stuff almost any cut, just for the fun of it. If it does not have a naturally occurring cavity, you can either butterfly it or cut a small pouch into its centre.

Stuffings

Bread stuffings, the kind that are trotted out every festive season, invariably find themselves at the epicentre of both happy memories and family feuds. We dare not intrude with much advice or novel concepts. From sliced brown to challah, bread stuffings are good for keeping the bird moist and sopping up those delicious roasting juices. They can be kept relatively straightforward, seasoned with leeks, celery, and herbs, or tarted up with everything from sausage and mushrooms to apples and chestnuts.

Grains, such as rice, farro, wheat berries, or couscous, also make great stuffings for poultry. They serve much the same purpose as a bread stuffing, soaking up the excess juice and fat from the bird as it roasts while keeping the bird itself moist.

Vegetables, such as cooked greens, leeks, or wild mushrooms, make attractive and flavourful stuffings. In most cases, you will want to cook the vegetables fully prior to stuffing, as they will not cook sufficiently inside a roast. Be sure to carve the roast to reveal a mosaic of stuffing and meat in each slice.

Fruits, chutneys, and *mostarde* add flavour and texture along with moisture to pork roasts. Roasted apples and quince, brandied prunes, chutneys, and *mostarde* can all be used to stuff a pork loin, blade roast, or shoulder.

Forcemeats and sausage make especially good

THE BEAST WITHIN

Stuffing one bird or beast inside another is nothing new. The ancient Romans did it, and so did the medieval French. The British fell in line and someone carried the concept to the Americas. Whether it is the classic Cajun *turducken* you crave or some other modern mythical megabeast, we implore you to think this one through before you begin. Although we are certainly not purists, and we do adore a meaty project, a large part of the joy of a roast, stuffed or otherwise, is the juxtaposition of crispy, brown exterior and moist, juicy interior. If a chicken and a duck are stuffed into a turkey, they are essentially steamed while the turkey roasts, which does keep the whole project moist but severely diminishes the ratio of crispy skin to moist meat. That is a trade-off we cannot endorse.

stuffings for leaner cuts and beasts, such as pork loin, crown roast, rabbit, and smaller poultry. They provide not just moisture but a good bit of added fat. They can also be added to bread or grain stuffings. Always be sure to take the internal temperature of a sausage stuffing. It must be cooked to at least 60°C to be safely eaten, regardless of the temperature of the meat.

How to Truss a Bird for Roasting

Trussing poultry yields a more consistent and compact shape, which helps the bird to cook more evenly and prevents the pieces that protrude from the body, such as the legs and wings, drying out during roasting.

1. **Season the bird thoroughly,** inside and out, then position, breast side up, on a work surface. Fold the wing tips behind the bird's back.

2. **Cut a piece of butcher's twine** at least 1 metre long. Place the middle of the twine behind the neck of the bird, then circle each end of the twine one full rotation around the wing joint on either side. Pull each end of the twine tightly in the direction of the legs; this will press the bottom of the wings directly against the body of the bird.

3. **Place the twine in the groove between the leg and the body,** then cross the twine directly underneath the bottom of the breastbone, pulling it tight as you would a shoelace.

4. **Cross the legs of the bird** and bring the ends of the twine around the bottom of each leg to tie the legs together. Circle the legs once with the twine, then reinforce this tie with a knot.

5. **Clip any extra twine away,** and voilà! You have a trussed, oven-ready bird.

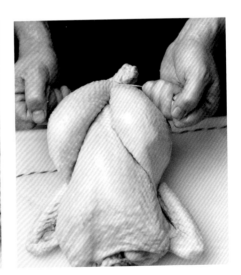

FIG-AND-SAUSAGE STUFFED QUAIL

We like to get our fill of fresh figs during their brief season in late summer. Roasted figs have an especially rich, honeyed flavour that is a good match for both poultry and pork. This stuffed quail manages to marry all three: a whole ripe fig is encased in an egg-shaped ball of pork sausage, which is then stuffed into the cavity of a brined quail. As the quail roasts, the sausage helps to keep the little bird moist and the hot fig exudes its sweet juices into the meat. **SERVES 4**

2 litres boiling water

130g fine sea salt

110g sugar

4 semi-boneless (glove-boned) quail (see page 67)

340g Lemon and Herb Pork Sausage (page 208)

4 small-to-medium figs, stems trimmed

1 tablespoon rendered duck fat or unsalted butter, melted

To make the brine, pour the boiling water into a large heatproof container or bowl. (Be sure to measure the water after it has come to a boil, as some will evaporate as it heats and measuring beforehand will result in overly salty brine). Add the salt and sugar and stir to dissolve. Let cool for several minutes then refrigerate until cold.

When the brine has cooled to 10°C, remove it from the fridge and add the quail. Make sure they are fully submerged, topping them with a plate if necessary. Cover the container, return it to the fridge, and leave the birds to brine for 3 hours.

Meanwhile, assemble the fig-stuffed sausage balls. Divide the sausage into four 85g portions. Shape each portion into a patty about 1cm thick. Place a fig, stemmed side down, directly on the centre of a sausage patty. Bring the sausage up around the fruit, moulding the sausage into a rough egg shape around the fig, with a narrower top and wider base. Repeat with the remaining figs.

Preheat the oven to 220°C (gas 7).

Remove the quail from the brine, drain them well and pat them dry. Gently pry apart the legs of 1 quail and stuff a sausage ball, narrow end first, into the cavity. Once the ball is snugly inside the quail, cross the bird's legs and tie them together with a 15cm length of butcher's twine. Repeat with the remaining quail.

Brush the quail on all sides with the duck fat and place them, breast side up, on a rack in a roasting tin, spacing them at least 5cm apart. Place the pan in the oven and roast for 22–24 minutes, until a thermometer inserted directly into the middle of the stuffing registers 60°C.

Remove the quail from the oven and let rest for 5 minutes, then snip the twine and serve.

PORK-SHOULDER POT ROAST STUFFED WITH GARLIC, GREENS, AND WALNUTS

Chock-full of greens, this simple pork shoulder pot roast, made with pork butt, makes a nourishing and comforting supper. Abundant, leafy Swiss chard tends to be available year-round and is the standard for this stuffing, but it is equally good made with spinach, mustard, kale, or other seasonal greens. You will need to prepare the roast the day before you plan to cook it. **SERVES 8-10**

1 whole boneless, skinless pork butt, about 4kg

Fine sea salt and freshly ground black pepper

3 bunches Swiss chard or other leafy greens, stemmed

10 cloves garlic, sliced paper-thin

85g chopped toasted walnuts

360ml pork, chicken, or duck broth (see Basic Rich Broth, page 44)

360ml dry red wine

One day ahead of cooking, season and ready the roast for stuffing. First, make the pocket for the stuffing by making a horizontal cut through the middle of the roast, following the seam where the bone was removed. Leave one of the four edges completely intact. Open the roast like a book. Season liberally on both sides with salt and pepper. Close the book, wrap tightly with cling film, and refrigerate overnight.

Remove the roast from the fridge and allow it to come to room temperature for 2 hours. Preheat the oven to 180°C (gas 4).

Bring a large pot of salted water to a rolling boil. Add the chard leaves and blanch for about 2 minutes. Drain and let cool, then squeeze out any excess water. Chop the chard coarsely.

Open the pork shoulder like a book, with the intact edge on your left. Arrange the chard over the bottom half of the roast in a neat layer, leaving a 2.5cm border uncovered surrounding it. Distribute the garlic evenly over the chard, followed by the walnuts. Fold the top half of the roast over the stuffing and tie tightly with butcher's

twine in three places, spacing the loops evenly and reinforcing the book shape.

Place a rack inside a large casserole dish. Place the pork shoulder, fatty side facing up, on the rack. (If you don't have a rack that fits your pot, halve a few leeks lengthwise, place them on the bottom of the pot, and put the roast on the leeks; they will support the roast nicely during cooking.)

Transfer the casserole to the oven and roast for about 45 minutes. Remove the dish from the oven and carefully pour off the rendered fat. Reserve these pan drippings for another use. Add the broth and wine to the pot and return it to the oven. Turn down the oven temperature to 150°C (gas 2) and continue to cook, basting the roast every 30 minutes, for about 2½ hours. The roast is ready when it is a rich golden brown, fork-tender, and a bit wobbly.

Transfer the roast to a cutting board and let it rest for 20 minutes. Snip the twine and cut into thick slices. Bathe each serving with a spoonful of the cooking juices.

WILD-MUSHROOM STUFFED PORK RIB ROAST

Versatile mushroom *duxelles* makes an impressive and wonderfully earthy stuffing for pork, elevating an everyday rib roast to special-occasion status. The *duxelles* flavours the pork and provides moisture for the lean muscle meat as it roasts, keeping it juicy and delicious. The addition of wobbly *gelée* to the *duxelles* gives the stuffing structure, binding it so that when it comes to the time to carve and serve, each slice has a fetching eye of wild mushroom. Prepare the roast the day before you plan to cook it. **SERVES 4–6**

5-rib pork loin rack with chine bone removed, preferably from the shoulder end
Fine sea salt and freshly ground black pepper
250g Wild Mushroom Duxelles (page 21)

One day ahead, butterfly, season, and stuff the roast. First, stand the rack on a cutting board so that the ribs point upwards and towards the right. Make an incision along the top of the roast, directly behind the ribs, pressing into the bone. Grip the loose strip of meat that results from this cut and peel the meat away from the bone as you carve against the ribs, to open the roast like a book.

Season liberally on both sides with salt and pepper. Stuff the *duxelles* into the pocket, leaving a 2.5cm border uncovered on all sides. Gently close the roast back up into its original shape, taking care not to displace the mushrooms. To tie the roast, lay it fat side down on your board and run a length of twine underneath it parallel to the ribs. Tie a butcher's knot (see pages 168–9) between the first two ribs of the rack. Repeat this knot between each rib to finish trussing, then wrap the roast in cling film and refrigerate overnight.

Remove the roast from the fridge and let it come to room temperature for about 30 minutes. Preheat the oven to 190°C (gas 5).

Place the roast, fat side up, on the rack of a roasting tin. Place the tin in the oven and roast for 30–40 minutes, until the fat has caramelised and turned a deep golden brown. Reduce the oven temperature to 150°C (gas 2) and continue to cook for 30–40 minutes, until a thermometer inserted into the thickest part of the loin registers 60°C.

Remove the roast from the oven and let rest for 20 minutes. Snip the twine and cut between the ribs to serve.

DUCK STUFFED WITH FARRO, FIGS, AND HAZELNUTS

Golden, crispy roast duck is a festive centrepiece for a celebratory meal, especially when stuffed with sage-scented sausage, red wine–soaked figs, toasted hazelnuts, and farro. Farro, an ancient variety of wheat popular in central Italy, is a favourite in the Fatted Calf kitchen where we add it to hearty soups and rustic salads. When cooked, it has a firm but chewy texture that also makes it perfect for use in stuffings. Although we like the toasty, nutty flavour of farro in combination with duck, this stuffing works equally well with quail or other birds. You will need to prepare the figs the day before you plan to cook your roast. **SERVES 6**

360ml dry red wine

2 tablespoons sugar

1 dried bay leaf

1 sprig thyme, plus 2 tablespoons chopped fresh thyme

^{1}/$_2$ teaspoon peppercorns

1 allspice berry

Fine sea salt and freshly ground pepper

12 dried figs

1 whole duck, 2–3kg

450g Breakfast Sausage (page 208)

330g cooked farro or wild rice

100g hazelnuts, toasted, skinned, and coarsely chopped

15g chopped fresh flat-leaf parsley

Prepare the figs one day in advance. In a small saucepan, stir together the red wine, sugar, bay leaf, thyme sprig, peppercorns, allspice, and a pinch of salt. Bring to a simmer over a high heat to dissolve the sugar and allow the flavour of the spices to bloom. Remove from the heat and pour over the figs. Cover and refrigerate overnight.

Bone the duck following the instructions for a Whole Boned Bird on pages 74–6, going around the legs and wings and making sure you do not make any holes in the skin. Work carefully around the breastbone, leaving the tenders attached to the breast. Trim any glands or blood vessels off the meat once it is completely off the bone. Season inside and out with salt and pepper.

Preheat the oven to 190°C (gas 5).

Drain the figs and cut off the stems, then quarter them lengthwise. In a bowl, combine the sausage, farro, figs, hazelnuts, chopped thyme, and parsley. Pat the stuffing into a cylinder about 8cm in diameter and 8cm shorter than the duck. Lay the stuffing directly on the middle of the duck and roll the meat tightly around it. Lightly score the skin in a crosshatch pattern to facilitate the release of fat during cooking. Truss the duck tightly (see page 183).

Place the duck on the rack of a roasting tin and transfer to the oven. Roast, basting after the first 20 minutes with the drippings that have accumulated in the tin, for about 1 hour or until a thermometer inserted into the middle of the stuffing registers 60°C. If the skin starts to get a little too brown, you can lower the oven temperature to 170°C (gas 3) so the duck finishes more slowly.

Remove the duck from the oven and let it rest for 10 minutes. Carve into slices 2.5cm thick to serve.

5

SAUSAGE,
SALAMI & THEIR COUSINS

Make sausage. Make sausage to reclaim the sweet memories of supper at *Nonna*'s, smoky backyard barbecues, and cold bonfire nights. Make sausage to experience once again the unabashed joy of tasting a plump Oktoberfest wurst. Make sausage to embark on the quest to capture the essence of an alluring lemongrass scent wafting in the air of a crowded night market. Grind, stuff, and link sausages with friends and family to discover where your food comes from and to appreciate the labour that brings it to your table. Make sausage and create something from practically nothing the way our ancestors have done for thousands of years.

Sausage making is a craft born of necessity, a way to make the most of the odd bits, scraps of meat, blood, fat, and entrails. At the most basic level, sausage is just seasoned ground meat. The ingredients and seasonings you use and the way the meat is prepared are what make each sausage unique.

Making your own sausage is a rewarding project that need not be complicated. If you are just starting to learn to make your own, begin with the basic sausages and their variations on pages 199–209. The methods for making simple fresh sausages are the basis for producing all types of sausage and salami. As you become more adept at marinating, grinding, and mixing, you might try your hand at casing fresh sausage. Once you have practised the various casing techniques, you will be able to crank out links and loops to prepare poached or smoked sausages, and if you want to go a step further, you can even naturally ferment your sausages to make salami.

The Sausage Shop

Although making sausage is not complicated, it does require space, time, tools, and a little forethought. Preparation is key. You will need to set up your own sausage shop or workspace, gather all the essential tools and ingredients, and have a thorough understanding of any recipe before you get cranking.

Workspace

You need plenty of room to cut, grind, and case meat. Choose a work area that affords a little elbow room, equipped with a work surface that is a comfortable height for you. Keep it free of inessentials. Have only the tools you need and a stack of clean kitchen towels. For easy access, lay your tools out on a baking tray or tray lined with a clean towel before you begin. If your end goal is a smoked sausage, dried, or salami, have ready a space to hang your finished sausage (see pages 232 and 234).

Time

Sausage takes time. Simple, uncased fresh sausages might take only an hour to assemble, but as you progress to cased and cooked sausages or dried sausage and salami, the process can take a few days. It's a good idea to break up the project into steps and spread these out over several days, rather than try to accomplish too much too quickly, especially while you are getting the hang of it. Plan to gather your materials a day or two ahead. Seasoning or marinating the meat is best done the day before, although two hours will suffice if you are pressed for time. If you are preparing a cooked sausage, you must allow an extra day to hang the fresh sausage before you poach or smoke it.

Tools of the Trade

Although you can make simple uncased sausage with freshly ground meat from a butcher, spices, and your own two hands, we encourage you to take the plunge and try your hand at grinding your own meat and casing your sausage. This requires a little investment in some good tools. If you are just beginning to make sausage, start simple and be realistic about what you actually need. You can always beef up your collection as you hone your craft.

Meat Grinders

Manual Grinder: This usually clamps to a table. You feed the cubes of meat into the hopper, turn the crank, and out comes freshly ground meat. This is great if you want to grind a relatively small quantity of meat for sausage patties, burgers, or meatballs. It's small and easy to assemble, clean, and store.

pusher

collar

cutter

feeder tube

scroll

screen

But for grinding more than a couple of kilos of meat, you may find it too time-consuming. Also, these old-fashioned grinders work well only if the blade is sharpened or replaced regularly.

Electric Multipurpose: If you already own a stand mixer, you can get a grinder attachment that is compact, easy to use, easy to clean, and affordable. The drawback is that these machines lack chutzpah. The blade and grinder screens are not as durable and effective as those on a larger machine. That means, to grind about 2.5kg of meat, you will need to allow about 10 minutes. But if you are just beginning, or if you only make sausage a few times a year and in small quantities, this type of machine will get the job done.

Electric Single Purpose: If you are a gearhead, a committed sausage fanatic, or looking to upgrade, you will want a stand-alone grinder. You'll find many different models of single-purpose, tabletop electric grinders on the market, and most of them will do the job well. Look for a machine that has electrical needs compatible with the voltage available in your workspace and parts that are easily replaceable.

Sausage Stuffers

Electric: If you already own a stand mixer with a grinder attachment, you can get a very affordable stuffer attachment that is compact and easy to use. Many manual and electric grinders also have optional stuffer attachments. The downside of a stuffer attached to a grinder is that the meat must pass through the mechanism a second time before it goes into a casing, which can negatively affect the texture by mashing the meat. However, if space and price are an issue, this is certainly an option.

Hand Crank: If you enjoy making sausage, a hand-crank stuffer is worth the investment. All of them

operate on the same premise. They are comprised of a base, a cylinder where you load the freshly ground sausage, and a crank or handle that you rotate to move the gears that control the piston that presses the sausage through the nozzle. Nozzles of different sizes work with assorted casings to produce sausages of varying diameters.

Hand-crank sausage stuffers come in horizontal and vertical designs. Horizontal stuffers, although often attractive, are not particularly practical. The design requires that they be positioned at the edge of a worktable, they often require two sets of hands to function properly, and they take up a considerable amount of space.

Vertical stuffers are the more common and versatile choice. A vertical stuffer requires only one user, has a crank located at the top of the machine that allows you to position it anywhere on the work surface, and takes up a minimum of surface area.

Smokehouses

Smokehouses can be created from virtually anything, from old oil drums and household chimneys to cinder blocks and plywood. If money is no object, you can buy deluxe models with temperature controls and digital displays. Most are bigger than a bread box, but some are the size of a walk-in wardrobe. If you are only planning to smoke one batch of sausages at a time, then you can get away with a smaller, bread-box-sized model.

When deciding what type of smokehouse is right for you, give yourself an honest appraisal of your DIY abilities and consider how frequently you will be using it; whether you will be using it to cold smoke, hot smoke, or both; how much you will be smoking at any given time; and how much room you have to store your smokehouse.

There is a discussion of smokehouses in Chapter 7, so if you're planning to make any of the smoked sausage recipes in this chapter, read pages 301–3 before you start.

Sausage Knife

A sausage knife has a small, sharp paring blade at one end and tines on the opposite end. It is a handy tool used during the stuffing process. The knife end is used to cut casings and twine, while the tines are used to pierce the filled casing to release air bubbles.

S Hooks

S hooks of various shapes and sizes are indispensable. Larger hooks (15–20cm) are great for hanging lengths of sausages to set overnight or in the smokehouse. Smaller hooks (about 2.5cm) are handy for hanging individual salami for long periods of time.

Scales

Most recipes for sausage provide measurements in weight. Not only is meat generally purchased by weight but a weight measurement is typically more accurate than a volume measurement.

A small set of digital electronic scales that measures in 10 gram increments will provide you with the most accurate results. Alternatively, if you want something a little more old school, a sturdy set of calibrated analogue platform scales with a capacity of at least 5kg will do the trick.

Ingredients: Meat and Fat

Sausage is comprised mainly of meat and fat, usually in a ratio of about 25 per cent fat to 75 per cent lean meat. When you shop for meat to make sausage, choose fattier or untrimmed cuts, that will provide you with a bit of both meat and fat.

Poultry and Rabbit

You can use nearly any type of poultry to make sausage. If you are using whole birds, bigger or more mature ones are better. The meat to bone ratio tends to be higher in a larger bird, and the muscles have had time to become more fully developed, yielding more flavourful results. If you have the option of using poultry parts, opt for breasts over legs and thighs, which have tendons that will need to be removed.

Larger rabbits are optimal for the same reasons. Except for the bony forelegs, you can use almost every part of the rabbit, including the hind legs, saddle, skirt, tenders, and kidneys. Even the liver can be ground in with the meat or diced and folded into the forcemeat.

Both poultry and rabbit are fairly lean and require additional fat. Using poultry fat to produce an all-poultry sausage is not recommended, however. Poultry fat is prone to smearing (see page 201) during grinding and liquefies readily during cooking, leaving you with a dry and unappetising sausage. Most commercially produced poultry sausages contain stabilisers such as collagen or gelatine to counteract smearing. Adding a little pork back fat is a natural alternative that works nicely.

Pork

The pig was practically designed for sausage making, and pork makes the most reliably delicious sausage. Good-quality pork is flavourful on its own, yet it can also play a good supporting role for a variety of other flavours.

Buy meat from a fully grown hog (it should be over 70kg). Mature hogs with well-developed muscles and a good fat content are preferred over younger, leaner pigs, which are better suited to roasting. The natural fat content of a boneless pork (collar) butt is usually ideal for sausage making, but you can also use a mixture of lean and fat trimmings from other cuts of pork, as long as the fat content is about 25 per cent. If you are using additional pork fat, harder back fat is preferable to the softer leg and belly fat.

Lamb

Lamb sausage has a distinctive flavour that stands up to intense spicing or strong, herbaceous seasonings. Boneless shoulder, shank, neck, and leg are all great cuts for sausage making. If needed, additional fat can be taken from the breast or shoulder.

Lamb fat can be very firm and needs to be ground twice. It can also be strongly flavoured. If you are making an all-lamb sausage, consider substituting olive oil for a portion of the fat or making a sausage with a fat content slightly lower than the traditional 25 per cent. Alternatively, if you tend to shy away from the stronger flavour of lamb, try using half pork and half lamb in your recipe for a sausage with a subtler lamb flavour and the great texture of pork.

Beef

Beef has an earthy flavour, and beef sausages tend to be highly spiced. Chuck, brisket, rump, and trimmings from roasts (such as the side muscle of the tenderloin) make excellent sausage. Brisket and chuck are usually at least 25 per cent fat, so no additional fat is needed. If possible, request untrimmed beef cuts from your butcher. If you are using leaner cuts and need additional fat, the rib-eye fat cap works well. Beef fat is firm, so always grind it twice. Avoid suet (kidney fat), as it is too soft for sausage making.

BASIC RECIPES FOR SAUSAGE

POULTRY OR RABBIT

2kg boneless, skinless poultry or rabbit
+ 500g pork back fat
+ 2 tablespoons fine sea salt

_____

   2.5kg basic poultry or rabbit sausage

LAMB

   1.5kg boneless lamb shoulder
+ 1kg lean boneless lamb foreshank or hind shank
+ 2 tablespoons sea salt
+ 2 tablespoons olive oil

_____

   2.5kg basic all-lamb sausage

Alternatively

   1.2kg boneless lamb shoulder
+ 1.2kg boneless pork picnic
+ 2 tablespoons sea salt

_____

   2.5kg basic lamb and pork sausage

PORK

   2.2kg boneless pork picnic
+ 225g pork back fat

   or

   2.5kg boneless pork (collar) butt
+ 2 tablespoons fine sea salt

_____

   2.5kg basic pork sausage

BEEF

   2.5kg untrimmed beef chuck or brisket
+ 2 tablespoons sea salt

_____

   2.5kg basic all-beef sausage

Alternatively

   1.5kg untrimmed beef such as chuck or brisket
+ 1kg pork (collar) butt
+ 2 tablespoons sea salt

_____

   2.5kg basic beef and pork sausage

Basic Sausage Method

The process for making nearly every kind of sausage begins with the same steps. First, you assemble the spices and cut the meat. Next, you mix the meat with the spices, leave it to marinate for a while, and then grind it. Once it is ground, the meat is mixed again by hand. The sausage is now ready to use or ready to case.

Step 1: Assemble the Spices

Your spices consists of the ingredients you will be using to flavour your sausage. Many sausage-making supply companies sell ready-made spice blends, but toasting and grinding your own spices makes a difference you can taste.

Begin by measuring the salt. Then measure your spices. If the recipe calls for toasted spices, toast them in a 170°C (gas 3) oven for 3–5 minutes. Allow them to cool, then grind them together in a spice grinder. For most sausage, unless otherwise indicated, grind your spices very finely. Mix the ground spices with the salt. If the recipe calls for garlic, mince it finely and add it to the spices along with any whole spices.

Step 2: Cutting

Cut the meat into relatively uniform cubes that are smaller than the opening of your grinder (for most grinders, 2.5cm cubes are best). Remove any blood vessels, tendons, or glands. Place the cubed meat in a non-reactive bowl or container large enough to allow room for mixing.

Step 3: Marinating

Evenly distribute half the contents of the spice blend over the meat. Using your hands mix well until evenly coated. Add the second half of the kit and mix again. Cover and refrigerate for at least 12 hours or up to 2 days to allow the seasonings to permeate the meat.

Step 4: Chilling

Sausage likes to be kept cold. Chilling both your meat and parts of the grinder helps to avoid grinding issues such as smearing (see page 201). Keeping the meat cold before and during the process also extends the shelf life of the finished sausage. After cutting and marinating the meat, be sure to refrigerate it for at least 2 hours and preferably overnight so that it is thoroughly chilled. Keep everything refrigerated until you are ready to grind.

Step 5: Grinding

Whichever type of grinder you use, the mechanics and setup are essentially the same. Begin by attaching the feeder tube to the base of the machine. Insert the scroll (see photo on page 195) into the tube. Attach the cutter, flat side out, to the scroll. Most grinders come with multiple screens to allow you to vary the size of the grind. Choose the screen for the type of grind you are trying to achieve and attach the it flush with the opening of the feeder tube. Screw the collar on to the end of the tube securely, but do not overtighten. If your grinder is equipped with a tray, attach it to the top of the feeder tube.

You will need a wide, non-reactive bowl or container that fits easily under the grinder to catch the ground meat. Take the meat from the fridge and feed it into the tube, one piece at a time. Let the machine do the work rather than push too much meat through at once. If you are using an electric grinder, allow the machine to run for a full minute after the last of the meat has been fed through the tube to expel any remnants. Wipe the face of the screen clean while the machine is still running and then turn the machine off.

In most cases, you will grind a batch of meat only once. The exceptions are burger meat, beef fat, lamb fat, and sausages with a very smooth consistency, which need to be ground twice.

Step 6: Mixing

Seasoned, ground sausage meat, known as forcemeat, needs to be mixed thoroughly by hand for 1–2 minutes. This action, similar to kneading bread dough, helps to develop the proteins that bind the sausage together. It also ensures that the seasonings are evenly distributed throughout. When a more homogenous texture is desired, some sausage meat is mixed further in a stand mixer fitted with a paddle attachment, or in a food processor. This process is called emulsifying.

Step 7: Tasting

Scoop up about 2 tablespoons of the well-mixed forcemeat and shape it into a small, flat patty. Cook the patty in a small pan over a medium heat. Evaluate the taste and texture. If the sausage seems dry and crumbly, incorporate a small amount of ground fat. If the seasoning needs to be more pronounced, add more salt or spices. If it is too highly seasoned for your taste, add a small amount of unseasoned ground meat and ground fat to help absorb some of the excess. Remember, it is much easier to add salt and spices than it is to lessen their intensity once the forcemeat is prepared. If you tend to like mildly seasoned sausage, start with about half the quantity of salt and spices and add more to taste if needed.

SMEARING

If the fat begins to squeeze out of the sides of the grinder in shiny, flat ribbons or through the die in greasy-looking streaks, stop! You have smearing, a condition that can ruin the texture of your sausage. You need to stop grinding, identify the cause, and remedy the situation. Here are three primary causes and their solutions:

1. The grinder or the meat is too warm. Check the temperature of the meat and the grinder. Wash the grinder and chill down the parts and any unground meat for 30 minutes, then start again.
2. The cutter is inserted backwards. Take the grinder apart. Wash and chill the parts and reassemble, making sure the cutter is facing flat side out.
3. The cutter blade is dull. Cutter blades do wear out over time. Keeping a spare blade on hand is always a good idea. Replace the blade and have the old blade sharpened.

THE UGLY BURGER

A burger is essentially a very basic beef sausage. Of course, now that we've stressed the importance of marinating and mixing, forget all of that. The guiding principle behind the Ugly Burger is to season the meat just before cooking and to manipulate it as little as possible for a homely but extraordinarily tender, juicy burger that will leave you wondering how something so ugly and simple could possibly taste so good. **MAKES 4**

450g boneless beef chuck, cut into 2.5cm cubes
450g boneless beef thick rib, cut into 2.5cm cubes
2 teaspoons fine sea salt

Place all the beef into a non-reactive bowl and refrigerate for 1 hour. Refrigerate the parts of your grinder until ready to use.

Prepare a hot barbecue.

Assemble the grinder with the medium screen and grind the chilled beef once. Sprinkle half the salt over the ground beef and grind again. Without mixing, divide the ground meat into 4 equal portions. Carefully flatten each portion into a patty 2.5cm thick and gently round the edges. A few cracks are okay. Make a small indentation in the middle of each patty with your thumb to keep the burger from contracting during cooking, then sprinkle the remaining salt over both sides.

Place the patties on the grill grate directly over the coals and grill, turning frequently, for about 4–6 minutes, or until done to your liking.

DUCK AND LEMONGRASS SAUSAGE PATTIES

These fragrant sausage patties are great grilled and used in a simple rice-bowl meal with herbs or in a *banh mi* (Vietnamese sandwich). The mixture can also be crumbled into a stir fry, grilled on skewers (or lemongrass, as in the photo), encased in a dumpling wrapper and steamed, or formed into little meatballs and combined with noodles in a noodle soup broth (page 45). **MAKES 8**

850g boneless, skinless duck meat (see
    pages 72–3), cut into 2.5cm cubes
280g pork back fat, cut into 2.5cm cubes
1 tablespoon fine sea salt
2 teaspoons freshly ground pepper
1 teaspoon sugar
2 teaspoons fish sauce
1 tablespoon minced shallot
1 tablespoon minced lemongrass
1¹/₂ teaspoons peeled, grated and chopped
    fresh ginger
1¹/₂ teaspoons minced garlic

Place the duck and fat in a non-reactive bowl or container. In another bowl, combine the salt, pepper, sugar, fish sauce, shallot, lemongrass, ginger, and garlic and mix well. Mix the spices into the meat (see page 200), cover, and refrigerate overnight.

Refrigerate the parts of your grinder until ready to use.

Following the instructions for grinding on page 200, fit the grinder with the smallest screen and grind the meat once. Mix the forcemeat well by hand for 2 minutes.

Cook a small sample of the mixture in a sauté pan and adjust the seasonings if necessary. For patties, divide the forcemeat into 4 equal portions and carefully flatten each one to make a patty 2cm thick. Grill, crumble, or otherwise enjoy as suggested above.

LAMB AND HERB MEATBALLS

Aleppo pepper, or *Pul biber*, is popular throughout the Middle East, the Mediterranean, and in our kitchen at the Fatted Calf, where it makes an appearance in numerous recipes, including many of our favourite lamb dishes. The vibrant, flaky ground chilli has a sharp, almost citrus-flavoured sting that is followed by a mellow, earthy current of heat, a quality that remains distinct even in the panorama of herbs and spices used to season these savoury meatballs.

These meatballs can be browned in olive oil and then simmered in Chilli Tomato Sauce (page 322), or skewered and grilled. Alternatively, to make lamb burgers, divide the forcemeat into 6 equal portions and shape each one into a patty 2cm thick. **MAKES ABOUT 36 SMALL MEATBALLS**

1 teaspoon fenugreek seeds

1 teaspoon black peppercorns

1 teaspoon coriander seeds

1 teaspoon yellow mustard seeds

1 large allspice berry

1 small dried bay leaf

1 tablespoon fine sea salt

1½ teaspoons Aleppo pepper flakes
   (pul biber)

2 teaspoons finely chopped garlic

1kg boneless lamb shoulder or leg, cut into
   2.5cm cubes

1 tablespoon grated yellow onion

2 tablespoons chopped fresh mint

2 tablespoons chopped fresh coriander

15g chopped fresh flat-leaf parsley

1 tablespoon olive oil

Preheat the oven to 170°C (gas 3). Spread the fenugreek, black peppercorns, coriander seeds, mustard, and allspice on a baking tray and toast for 3–5 minutes, until fragrant. Let cool completely, then transfer to a spice grinder, add the bay, and grind finely.

In a small bowl, combine the freshly ground spices, salt, Aleppo pepper flakes (*pul biber*), and garlic and mix well. Place the lamb in a non-reactive bowl or container. Mix the spices evenly with the meat (see page 200), cover, and refrigerate overnight.

Refrigerate the parts of your grinder until ready to use. Following the instructions for grinding on page 200, fit the grinder with the smallest screen and grind the meat twice. Add the onion, mint, coriander, parsley, and olive oil and mix well by hand for 2–3 minutes. The mixture should begin to firm up and feel cohesive. Cook a small sample of the mixture in a sauté pan and adjust the seasonings if necessary. Roll the mixture into meatballs about 4cm in diameter, then use in one of the delicious ways specified above.

OAXACAN-STYLE CHORIZO

This is a boldly seasoned chorizo we developed for our friends at Rancho Gordo, in Napa, California, when they were hosting esteemed Mexican-food writer Diana Kennedy. In addition to selling a wide variety of dried beans, chillies, grains, and other food products, Rancho Gordo carries an intensely tropical banana vinegar that provides a fruity counterpoint to the earthy, fiery chillies.

We usually leave this chorizo loose for use in little tacos, for mixing with potatoes, or for adding to soups and stews. It can also be put into pork casings about 30cm long, tied into loops, hung overnight in a cool place, and then grilled slowly or smoked. **MAKES A GENEROUS 2.5KG**

2kg boneless pork picnic, cut into 2.5cm cubes

450g pork back fat, cut into 2.5cm cubes

40g fine sea salt

6 guajillo chillies

6 ancho chillies

2 chipotle chillies

3 dried chiles de árbol

2 teaspoons black peppercorns

6 allspice berries

2 tablespoons cumin seeds

4 whole cloves

4cm piece of cinnamon stick

5 cloves garlic, unpeeled

125ml fruit vinegar (such as banana, pineapple, or cider)

1 teaspoon annatto (achiote) powder

2 tablespoons chopped fresh thyme

2 tablespoons chopped fresh oregano

Place the meat and fat in a large non-reactive bowl. Season the meat with the salt. Cover and refrigerate overnight. Refrigerate the parts of your grinder until ready to use.

Toast all the chillies in a heavy-based frying pan over a medium heat for about 1 minute, until fragrant and pliable. Turn out on to a plate to cool. Seed the chillies and discard the seeds and stem, then tear the chillies into smaller pieces. Place them in a bowl, add warm water to cover, and let soak for about 30 minutes, until softened. Drain, reserving the chillies and a few tablespoons of the soaking liquid separately.

Using the same pan, toast the peppercorns, allspice, cumin, cloves, and cinnamon stick over a medium heat for 2–3 minutes, until fragrant. Let cool to room temperature, then transfer to a spice grinder and grind finely.

Dry roast the garlic for about 5 minutes in the same pan over a medium heat, until the skin darkens and the garlic releases a fragrant aroma. When the cloves are cool enough to handle, peel and chop.

In a blender, combine the vinegar, chillies, ground spices, garlic, and annatto powder and purée until smooth. If the mixture is too thick, thin as needed with the reserved chilli-soaking liquid to achieve a good consistency. Scrape the purée into a large non-reactive bowl and fold in the thyme and oregano. Place the bowl directly under the meat grinder.

Following the instructions for grinding on page 200, fit the grinder with the largest screen and grind the meat once directly into the chilli-spice purée. Mix the forcemeat well by hand for 2–3 minutes. It should begin to firm up and feel cohesive. Cook a small sample of the mixture in a sauté pan and adjust the seasonings if necessary. Use as suggested above.

SAUSAGE AND SEASONING CHART

You can use the basic sausage recipes (page 199) to create a variety of different sausages by altering the spice blend.

Sausage	Basic Sausage Recipe	Spice Blend	Grind	Garnish	Casing
Basque	Pork	1 teaspoon freshly ground black pepper 2 tablespoons piment d'Espelette 1 teaspoon toasted and ground aniseeds 1 dried bay leaf, ground 1 tablespoon minced garlic	Medium	60ml dry red wine	Lamb
Black Truffle	Pork	1/2 teaspoon freshly ground black pepper 3 allspice berries, ground 1 teaspoon minced garlic	Fine	30g fresh black truffle, grated 2 tablespoons Cognac	Lamb or Pork
Bordelaise	Pork	1 1/2 teaspoons freshly ground black pepper 2 tablespoons minced garlic	Fine	360ml dry red wine reduced to 120ml 15g chopped fresh flat-leaf parsley	Lamb
Breakfast	Pork	1 1/2 teaspoons freshly ground black pepper 1 teaspoon ground cayenne pepper 1/8 teaspoon freshly grated nutmeg 1 allspice berry, ground 1 tablespoon minced garlic	Medium	30g finely chopped fresh sage	Lamb
Calabrese	Pork	25g toasted and ground fennel seeds 2 tablespoons toasted and ground cumin seeds 2 teaspoons freshly ground black pepper 1 teaspoon freshly ground white pepper 2 tablespoons ground chilli flakes 1 1/2 teaspoons ground dried oregano 2 tablespoons minced garlic	Fine	60ml dry white wine	Pork
Farmer's Sausage	Pork	1 1/2 teaspoons freshly ground black pepper 1 teaspoon toasted and ground aniseeds 1 teaspoon ground cayenne pepper	Medium	360ml dry red wine reduced to 120ml 2 tablespoons grated Pecorino Romano cheese	Pork
Fennel	Pork	2 tablespoons toasted and ground fennel seeds 1 1/2 teaspoons freshly ground black pepper 2 teaspoons ground dried oregano 4 teaspoons minced garlic	Medium	30g chopped fresh flat-leaf parsley 2 tablespoons whole fennel seeds, toasted 1 tablespoon chilli flakes 60ml dry white wine	Pork
Lemon and Herb	Pork	1 1/2 teaspoons toasted and ground coriander seeds 2 dried bay leaves, ground	Fine	Grated zest of 2 lemons 60ml dry white wine 30g chopped fresh herbs, such as flat-leaf parsley, chives, thyme, oregano, or sage	Lamb

IN THE CHARCUTERIE

Sausage	Basic Sausage Recipe	Spice Blend	Grind	Garnish	Casing
Merguez	Lamb	1½ teaspoons toasted and ground cumin seeds 1½ teaspoons toasted and ground coriander seeds 1 teaspoon freshly ground black pepper 1 teaspoon toasted and ground fennel seeds 1 star anise, toasted and ground 2 allspice berries, toasted and ground 1 tablespoon plus ½ teaspoon paprika ⅛ teaspoon ground cinnamon 2 tablespoons ground cayenne pepper	Fine	Grated zest of ½ lemon mixed with 1 tablespoon olive oil	Lamb
Rabbit Boudin	Rabbit or Pork	1 teaspoon freshly ground white pepper 1½ teaspoons ground yellow mustard seeds 1 tablespoon ground dried thyme 1½ teaspoons piment d'Espelette	Fine, twice	None	Lamb
Spicy Italian	Pork	3 tablespoons toasted and ground fennel seeds 1 teaspoon freshly ground black pepper ½ teaspoon freshly ground white pepper 4 teaspoons ground chilli flakes 1 tablespoon plus ½ teaspoon paprika 1 tablespoon minced garlic	Medium	2 tablespoons dry white wine	Pork
Sweet Italian	Pork	3 tablespoons toasted and ground fennel seeds 1½ teaspoons toasted and ground aniseeds 3 allspice berries, ground 1 teaspoon freshly ground pepper 1 tablespoon ground dried oregano 1 tablespoon minced garlic	Medium	30g chopped fresh flat-leaf parsley 60ml dry white wine	Pork
Toulouse	Pork	1½ teaspoons freshly ground black pepper 3 whole cloves, ground ⅛ teaspoon freshly grated nutmeg 5 allspice berries, ground 2 tablespoons minced garlic	Medium	None	Pork
Wild Mushroom	Pork	1½ teaspoons freshly ground black pepper 2 allspice berries, ground 1 dried bay leaf, ground 1 teaspoon sweet (dulce) pimentón	Fine	165g sautéed wild mushrooms 2 tablespoons chopped fresh thyme 2 tablespoons chopped fresh flat-leaf parsley 60ml dry red wine	Lamb

Stuffed Sausage

Stuffing sausage has traditionally been the natural outcome of practical butchery. 'Waste not, want not' was the mantra of our ancestors, and so all of the random bits and pieces were seasoned, ground, and stuffed into the animal's intestines. If you are getting serious about your sausage, you will want to try your hand at the time-honoured craft of casing sausage. Stuffing your sausage takes it to the next level. Loops, coils, links of varying lengths, and precious packets wrapped in webbed caul fat are the hallmarks of sophisticated butchers and charcutiers. With a little practice, you can crank out an array of meaty beauties and continue the age-old tradition.

Sausage Casings

Natural casings are the processed lamb, pork, and beef intestines used for casing sausage and salami. Somewhat permeable, they allow the perfume of smoke or the flavour of a poaching liquor to penetrate the meat while their contents remain juicy and secure. Natural casings come in a variety of different diameters to suit your needs and can be purchased from a speciality butcher or sausage supply company.

The lamb, pork, and beef casings are typically sold packed in salt or brine. Before they can be used, they must be rinsed thoroughly in several changes of cold water. Once rinsed, they can be stored immersed in water in the fridge for up to 5 days. For longer storage, drain them, pack them in fine sea salt, and refrigerate them for up to 6 months.

The casings that follow are the most commonly used in the charcuterie, and the ones used throughout the recipes of this book.

Lamb Casings: Lamb casings are the most delicate natural intestinal casings, and they tend to require a little more practice and attention. They are usually sold in two different diameters: 22–24mm, used for breakfast links or *merguez*, and 24–26mm, also known as a frankfurter casing, commonly used for hot dogs. Lamb casings are sold by the hank (80m), but you can often purchase smaller quantities from a good butcher. You will need about 1 metre of lamb casing for every 450g of meat.

Hog Casings: Hog or pork casings are the most common, versatile, and easy-to-use natural casings. They are traditionally used for many different types of sausages, including most Italian sausages, *kielbasa*, and *boudin noir*. Hog casings come in a variety of sizes. The 32–35mm size is a good all-purpose casing for most sausages. They are sold by the hank (80–90m), but you can usually buy smaller quantities from a good butcher. You will need about 60cm of hog casing for every 450g of meat.

Beef (Ox) Middles: Beef middles, or runners, are a natural part of the beef intestine traditionally used for salami and *cotechino*. They are generally sold by the piece and can be as long as 18m. Plan on using about 30cm of beef middle for every 450g of sausage.

Sewn beef middles have been reinforced and closed at one end, making them very sturdy and perfect for medium-sized cured salami. Like the regular middles, they are sold by the piece, but they are much more expensive.

Beef middles come packed in salt and need to be thoroughly rinsed in several changes of water then soaked overnight. Be sure to turn the middle inside out before using to expose its smooth interior. Once rinsed, it can be stored in water in the fridge for up to 5 days. For longer storage, middles should be drained, salted, and frozen.

Beef Caps: Beef caps, or bungs, are another part of the natural beef intestine. Much larger in diameter than middles, they are the traditional choice for haggis, *mortadella*, and other large salami. They are usually sold by the piece, and though they range in size, most caps can hold about 3kg of meat.

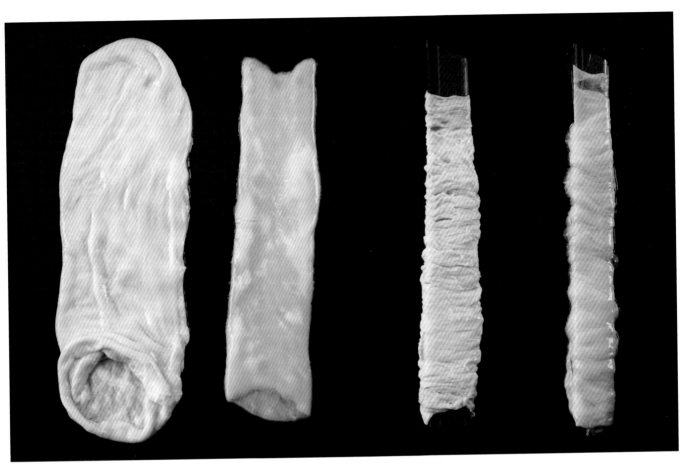

From left: beef cap/bung, beef middle, hog casing, lamb casing

Beef caps come packed in salt and need to be thoroughly rinsed in several changes of water then soaked overnight. As with beef middles, you should turn caps inside out prior to use to expose their smooth interior. Once rinsed, they can be stored in water in the fridge for up to 5 days. For longer storage, caps should be drained, salted, and frozen.

Caul Fat: Caul fat is the lacy, weblike fat harvested from around the pork liver that is traditionally used to line terrines or wrap *crépinettes* and other uncased sausages into neat packets. Extremely sturdy and versatile, it can be purchased fresh or frozen. Unlike intestine casings, it is not sold salted, but it must still be well rinsed to rid it of any remaining blood and then squeezed dry before use. You will need about 450g of caul fat to wrap 1.5–1.8kg of meat.

Any unused caul fat can be stored immersed in water with a little white vinegar for up to 5 days. For longer storage, squeeze out as much water as possible and freeze.

Stuffing and Linking Sausage

Stuffing and linking take practice. Don't be discouraged if your first few batches of sausage look a little irregular. They will still taste great. With practice, your moves will improve and, before you know it, you'll be cranking like a pro.

Assembling the Stuffer

1. **Set the base on a work surface.** A damp towel or C-clamp will help to keep the stuffer in place as you work.

2. **Screw the lid on to the piston.**

3. **Attach the handle** and rotate it until the lid ascends to its topmost position.

4. **Transfer the sausage meat into the cylinder** using your hands, packing it down firmly each time.

5. **Set the cylinder into the base.**

6. **Place a baking tray underneath the nozzle** to hold the finished sausage as you stuff. You are now ready to begin.

BLOWOUTS

Blowouts happen. Even people who have been making sausage for years experience the exploding-sausage phenomenon. If you don't bust the occasional casing, you are probably not filling it tightly enough. When a casing does burst, simply squeeze out the forcemeat and set it aside. Cut away and discard the damaged portion of the casing, slip the new end over the nozzle, and continue to case. Reload the reserved forcemeat into the stuffer at the next opportunity; make sure the casing isn't knotted at the end of the nozzle to avoid trapping air as you reload the cylinder.

Stuffing

There will always be a small amount of sausage left in the bottom of the cylinder and in the nozzle that won't make it into the casing. This extra forcemeat can be rolled into little meatballs, flattened into patties and grilled like burgers, added to soups, or used as a stuffing for vegetables or roasts.

1. **Remove a length of soaked casing from the water.** Open one end and pour about 1 tablespoon of water into the casing to lubricate it. Thread the entire length of casing on to the nozzle.

2. **Turn the crank so that the lid presses gently on to the top of the meat,** forcing just 1cm of the meat out through the nozzle. This helps to eliminate air pockets.

3. **Pull the end of the casing over the edge of the nozzle, then knot the end of the casing.** (If you are using sewn beef middles or beef caps, skip the knot, as they are already closed at the end.)

4. **Place your thumb and forefinger around the end of the nozzle to regulate the movement of the casing.**

5. **Crank the handle slowly to press the sausage meat into the casing.** Release more casing off the nozzle as the sausage flows through the tube.

6. **If an air bubble forms, prick the sausage casing with the tines of a sausage knife to release it.** If you begin to run out of casing, pause. Leave yourself at least 8cm of unstuffed casing and remove the length from the nozzle. Continue with a new length.

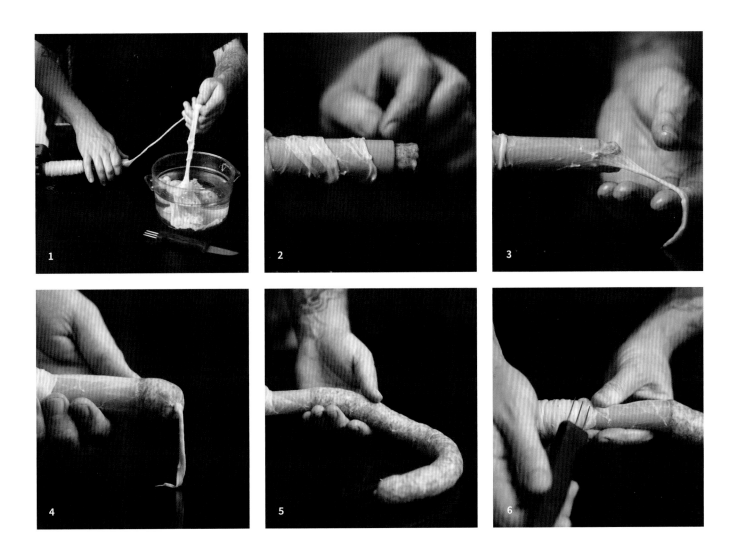

Linking

Once all of your sausage is cased, you can begin the process of linking. You can make links in both hog and lamb casings in a variety of sizes.

1. **Decide the length of your sausages.** Cut a length of twine to the desired length of your individual sausage links to use as a guide.

2. **Use the guide to pinch off your first length of sausage.** Align the string with the knotted end of stuffed casing and pinch down with your thumb and forefinger at the other end of the string to separate the first link from the stuffed casing.

3. **Twist to secure the link.** Rotate the link *away* from you about seven times.

4. **Start a second link.** Move the string along to align with the end of the first link.

5. **Using the string as a guide, pinch off the second link.** Pinch down at the end of the second link, then rotate *towards* you about seven times.

6. **Continue linking.** Repeat this process alternating the direction of rotation with each link.

7. **Separate the links.** When all of the sausage has been cased, snip between each link with scissors leaving a little 'tail' on either end of each sausage to prevent the forcemeat from spilling out during cooking. If you will be hanging your sausage overnight, keep the links attached in lengths of four to eight, and hang each length in a refrigerated location.

Coiling

Instead of linking the sausage, you can form impressive coils that make easy work of grilling. Thinner coils are made with lamb casing, and somewhat thicker coils are made with hog casing. To set the coils, you will need metal or bamboo skewers that are at least 25cm long. If you are using bamboo skewers, soak them in cool water for at least 30 minutes.

1. **Link the sausage.** Follow the directions for making links (see opposite), but pinch off and rotate lengths of 45–60cm, then snip apart with scissors as directed.

2. **Start coiling.** With one hand, press one tail end of casing on to your work surface to create a centre point and hold it firmly. Then, with the other hand, tightly wind the sausage around the centre point.

3. **Secure the coil.** Pierce the outer end of the casing (about 1cm from the tail of the link) with the pointed end of a skewer and insert it directly through the centre of the coil, parallel to the work surface.

4. **Skewer again.** Rotate the coil 90 degrees and run a second skewer directly through the centre of the coil.

Looping

Sausage that you will be hanging for smoking or air-drying is often cased in loops, which work best if hog casings have been used.

1. **Prepare to case your sausage.** Pull about 2.5cm of the end of the casing over the opening of the nozzle and knot.

2. **Cut a 12cm length of butcher's twine.** Tie it directly below the knot.

3. **Stuff the sausage.** Crank the handle slowly to press the sausage meat into the casing to form an 45cm length. You will want to stuff links to be looped more fully and firmly than you would regular links.

4. **Loop the sausage.** With one end still attached to the nozzle, form the link into a loop.

5. **Tie.** Tie the two ends together with the loose ends of the twine, then reinforce with a second knot.

6. **Loop the twine.** Create a loop of twine 2.5cm in diameter by knotting the two loose ends of twine. You will use this twine loop to hang your looped sausage.

7. **Start a second loop.** Pull 5cm of empty casing off of the nozzle and tie a knot with a new 10cm length of twine to begin the next loop. Then, using scissors, sever the first loop.

 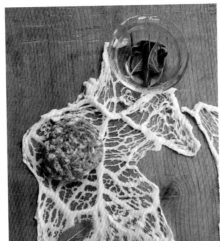 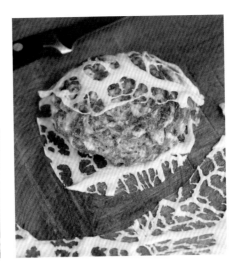

Wrapping

Wrapping sausage in caul fat is a great alternative to using casings, especially if you are making a small amount for fresh consumption or do not have access to a sausage stuffer.

1. **Divide the sausage.** Using scales, divide the loose sausage into equal-sized portions. In general, 115–140g is a good portion size.

2. **Shape the sausage meat** by hand into patties 2cm thick, smoothing the edges and eliminating any cracks.

3. **Prep the caul fat.** Remove a piece of caul fat from the water and squeeze well to remove any excess. Spread the fat out on a work surface.

4. **Lay the sausage patties on the caul fat,** leaving a 5cm border of fat around each patty.

5. **Wrap.** Pull the edge of the caul fat over the patties. Trim any excess, then pinch to close the seam.

Cooked Sausage

Many traditional sausages are finished with an extra cooking step, such as poaching or smoking. From rich blood sausage to smoky *kolbász*, delicate *bockwurst* to snappy hot links, the process of cooking is used to impart flavour, provide texture, and to help preserve the sausage.

Poaching

Poaching is used to set or firm the texture of a sausage with a wet or soft forcemeat, a step that allows it to be cooked effortlessly at another point. Some sausages are poached immediately following casing; others are better if they are poached after being hung overnight.

Heat a large pot of water before you begin grinding your meat. An 11-litre stockpot will comfortably accommodate a 2.5kg batch of sausage. Do not let the water come to a boil; instead, keep it just below a simmer. Season the water with salt. For every 4 litres of water, add 3 tablespoons of fine sea salt.

Grind, case, and link your sausage, but leave the links attached rather than clipping them apart. Insert a thermometer to check the temperature of the poaching water. The optimal is about 70°C.

Transfer the links to the pot and allow them to poach gently. Meanwhile, prepare a large ice-water bath. To ensure the links cook evenly, carefully stir them every now and again. The total cooking time is usually 15–20 minutes. The sausages are ready when a thermometer inserted into the centre of a sausage registers 60°C. Using a slotted spoon, remove the links and place them directly into the ice-water bath to stop the cooking and help to prevent shrinkage.

When the links are thoroughly chilled, remove them from the ice bath and clip them into individual links. Line a baking tray or tray with a clean kitchen towel and place the links on top. Refrigerate, uncovered, overnight. The next day, transfer the sausages to an airtight container and return them to the fridge. Storage time varies from recipe to recipe, but is usually around 5 days.

THE SAUSAGE PARTY

Inviting over a few game friends to join you in making sausage can lighten the workload and make the process even more fun.

Ahead of time, decide what types of sausage you want to make. If this is the first time for some people, choose just one type of stuffed sausage, or opt for a fresh sausage and *crépinettes* or patties. If you are old pros, choose a few different recipes, including something that you will poach or smoke. Decide who will be responsible for purchasing and gathering specific supplies, and mail order any special items, such as casings or curing salt, if necessary. Make sure you have all the necessary equipment and that it is in good working order.

One day before you plan to meet, gather your supplies. Shop for your meat and any other ingredients you will need. Prepare your spices, then cut your meat and marinate it.

Get together for the main event of grinding and casing.

BOCKWURST

Bockwurst is a classic German sausage made from a mixture of veal and pork and seasoned with fresh herbs. Traditionally, it was eaten in the spring to accompany the rich, malty bock beer that had been brewed by monks in the winter months for drinking at the end of Lenten fasting. Although you can enjoy this wurst any time of year, a few links browned in butter or cooked on the grill and served alongside green garlic mashed potatoes and a tall glass of beer seems like the perfect welcome to the spring season. **MAKES ABOUT 2.5KG OR 18–20 (15CM) LINKS**

1kg boneless veal shoulder, cut into 2.5cm cubes

800g boneless pork picnic, cut into 2.5cm cubes

340g pork back fat, cut into 2.5cm cubes

1/2 teaspoon white peppercorns

4 whole cloves

1/4 teaspoon mace blades

3 allspice berries

40g fine sea salt

1/4 teaspoon ground ginger

GARNISH

3 eggs

240ml heavy cream

25g thinly sliced fresh chives

15g chopped fresh sage

Finely chopped zest of 1 lemon

10 litres water seasoned with 140g fine sea salt for poaching

3m prepared hog casings (see page 210)

Place the veal, pork, and fat in a large non-reactive bowl or container. In a spice grinder, combine the pepper, cloves, mace, and allspice and grind finely. Transfer to a small bowl and stir in the salt and ginger. Mix the spices evenly with the meat (see page 200), cover, and refrigerate overnight.

To assemble the garnish, in a bowl, whisk together the eggs and cream until just blended. Fold in the chives, sage, and lemon zest. Refrigerate until ready to use.

Refrigerate the parts of your grinder until ready to use. Following the instructions for grinding on page 200, fit the grinder with the smallest screen and grind the meats twice. Pour in the garnish and mix well by hand for 3 minutes. The forcemeat needs to be mixed especially well to keep the cream from separating during cooking. Cook a small sample of the mixture in a sauté pan and adjust the seasonings if necessary.

Fill a large stockpot with the salted poaching water and heat on the hob to 70°C.

Following the instructions on pages 212–214, stuff the forcemeat into the hog casings and link into sausages about 15cm long, leaving the sausages connected.

Place the links in the poaching water. It may be necessary to adjust the heat, as the temperature will drop slightly when you add the links to the pot. Poach them slowly for 15–20 minutes, until a thermometer inserted into the centre of a sausage registers 60°C. Meanwhile, prepare a large ice-water bath and line a baking tray with a clean kitchen towel.

When the links are ready, transfer them to the ice-water bath for about 20 minutes, until well chilled. Drain the links, clip them apart, and arrange them on the prepared tray. Refrigerate, uncovered, overnight.

The next day, wrap the sausages tightly in cling film and refrigerate for up to 5 days, or wrap carefully and freeze for up to 3 months.

BLOOD SAUSAGE WITH CARAMELISED APPLES AND COGNAC

Sweet caramelised apples and slowly cooked onions combined with the punch of Cognac and the earthiness of pork and pork blood make a superb link that will win over blood-sausage sceptics. The poached links can be grilled, sautéed with apples, potatoes, and onions, or sliced and tucked between the layers of a root vegetable gratin. **MAKES ABOUT 3.5KG OR 22–24 (15CM) LINKS**

225g sliced onions

1 tablespoon unsalted butter

60g fine sea salt

2kg fatty pork butt, cut into 2.5cm cubes

2 teaspoons peppercorns

3 dried bay leaves

3 whole cloves

6 allspice berries

2 teaspoons piment d'Espelette

GARNISH

60ml lard or leaf lard

450g peeled and diced tart apples

1½ teaspoons fine sea salt

15g chopped fresh thyme (stems reserved)

120ml Cognac

1 litre pork blood, well chilled

350g pork back fat, cut into 0.5cm cubes

PANADE

30g fresh breadcrumbs

240ml heavy cream

COURT BOUILLON

10 litres water

140g fine sea salt

½ onion, sliced

1½ teaspoons peppercorns

5 dried bay leaves

Reserved thyme stems

4.5m prepared hog casings (see page 210)

Slowly sweat the onions, butter and ½ teaspoon of the salt, in a sauté pan over a low heat for about 20 minutes, until tender and translucent. Let cool.

Place the pork in a large non-reactive bowl or container. Finely grind the peppercorns, bay, cloves, and allspice berries in a spice grinder. Transfer to a small bowl and stir in the *piment d'Espelette* and the remaining salt. Mix the spices and cooled onions evenly with the meat (see page 200), cover, and refrigerate overnight.

Melt the lard in a large sauté pan over a medium-high heat. When it begins to sizzle, add the apples and cook, stirring occasionally, for about 10 minutes until they are a rich golden brown. Add the salt and thyme and cook for 1 minute longer. Remove from the heat, pour in the Cognac, and stir with a wooden spoon to loosen the fond from the bottom of the pan. Set aside to cool.

To make the *panade*, mix together the breadcrumbs and cream. To make the court bouillon, combine all the ingredients in a large stockpot over a low heat, and heat to 70°C.

Refrigerate the parts of your grinder until ready to use. Following the instructions on page 200, fit the grinder with the medium screen and grind the meat once. Pour in the pork blood, then the *panade* and mix well. Fold in the back fat and apples with their cooking juices. Mix the forcemeat by hand, using a quick, eggbeater-like rotating motion, for about 2 minutes, until it holds together and tightens somewhat. This forcemeat will be a bit looser than other sausage forcemeats. Cook a small sample of the mixture in a sauté pan and taste it. It should be rich, well seasoned, and a tad spicy, with plenty of flavour from the apple and Cognac.

Following the instructions on pages 212–214, stuff the forcemeat in the hog casings and link into sausages about 15cm long, leaving them connected. Place the links in the court bouillon. Check the heat as the temperature will drop slightly when you add the links to the pot. Poach slowly for 25–30 minutes, until a thermometer inserted into the centre of a sausage registers 65°C. Meanwhile, prepare a large ice-water bath and line a baking tray with a clean kitchen towel.

Cool the links in the ice bath for about 20 minutes, until well chilled. Drain, clip, and arrange the sausages on the prepared tray. Refrigerate, uncovered, overnight. The next day, transfer to an airtight container and refrigerate for up to 5 days, or wrap carefully and freeze for up to 2 months.

IN THE CHARCUTERIE

COTECHINO

When winter is in the air and the holidays are upon us, *cotechino* makes its seasonal appearance on the chalkboard at the Fatted Calf. A speciality of northern Italy, *cotechino* has a unique, springy texture due to the unusual addition of cooked pork skin. Although this large, lightly fermented sausage is delicious all year round, it is traditionally added to winter favourites such as *Bollito Misto* (page 117) or simmered with lentils (as pictured) and eaten for good luck on New Year's Day. (If you are serving the sausages with lentils, add the lentils and any vegetables or herbs to the pot about midway through cooking.) For a truly festive holiday variation on the classic *cotechino*, grate 30g fresh black truffle and add it to the forcemeat after grinding. **MAKES ABOUT 2.5KG OR 5 (450G) SAUSAGES**

600g pork skin, cut into strips 2.5cm wide

2 teaspoons coriander seeds

4 allspice berries

1¹/₂ teaspoons black peppercorns

1 teaspoon white peppercorns

2kg pork butt, cut into 2.5cm cubes

40g fine sea salt

1 teaspoon curing salt no. 1

¹/₈ teaspoon ground cinnamon

1 teaspoon ground ginger

60ml dry white wine

1.5m prepared beef middles, turned inside out (see page 210)

Bring a large stockpot filled with water to a rolling boil. Add the pork skin and cook for 10 minutes. Drain and pat dry with paper towels, then transfer to a bowl, cover, and refrigerate for 1 hour.

Preheat the oven to 170°C (gas 3). Spread the coriander seeds on a baking tray and toast for 3–5 minutes, until fragrant. Let cool completely, then transfer to a spice grinder, add the allspice and black and white peppercorns, and grind finely.

Place the pork in a large non-reactive bowl or container. In a small bowl, combine the freshly ground spices, sea salt, curing salt, cinnamon, and ginger and mix well. Mix the spices and the chilled pork skin evenly with the meat (see page 200), cover, and refrigerate overnight.

Refrigerate the parts of your grinder until ready to use. Following the instructions for grinding on page 200, fit the grinder with the medium screen and grind the meat once. Pour in the wine and mix the forcemeat by hand for 2 minutes, until it firms and holds together. Following the instructions for salami (see page 232), stuff the forcemeat into the beef middles. Hang the sausages in a cool, dry room for 3 days, then wrap loosely in a kitchen towel and refrigerate. They will keep for up to 1 week.

To cook the sausages, fill a large stockpot with lightly salted water and heat to about 75°C. Add the sausages and simmer for 2 hours.

BELGIAN BEER SAUSAGE

A rich, delicately hopped Belgian-style amber or blonde ale is key to the success of this sausage. Belgian-style beers tend to have a creamy froth, almost tropical aroma, and subtle flavours of ginger, coriander, and caramel that marry delectably with pork. Grill the finished links and top with caramelised onions or brown in butter, then simmer with Traditional Sauerkraut (page 319) and a splash of beer. **MAKES ABOUT 2.5KG, OR 18–20 (18CM) LINKS**

1½ teaspoons black peppercorns

2 teaspoons yellow mustard seeds

2 teaspoons coriander seeds

3 allspice berries

40g fine sea salt

1 teaspoon curing salt no. 1

2 teaspoons light brown sugar

½ teaspoon ground mace

½ teaspoon ground cayenne pepper

¼ teaspoon ground ginger

1 tablespoon minced garlic

¼ teaspoon finely chopped orange zest

1.7kg boneless pork picnic, cut into 2.5cm cubes

570g pork back fat, cut into 2.5cm cubes

300ml Belgian-style beer, well chilled

10 litres water seasoned with 140g fine sea salt for poaching

3m prepared hog casings (see page 210)

Preheat the oven to 170°C (gas 3).

Spread the peppercorns, mustard and coriander seeds, and allspice on a baking tray and toast for 3–5 minutes, until fragrant. Let cool completely, then transfer to a spice grinder and grind finely.

In a small bowl, combine the freshly ground spices, sea salt, curing salt, brown sugar, mace, cayenne, ginger, garlic, and orange zest and mix well. Place the pork and fat in a large non-reactive bowl. Mix the spices evenly with the meat (see page 200), cover, and refrigerate overnight.

Refrigerate the parts of your grinder until ready to use. Following the instructions for grinding on page 200, fit the grinder with the smallest screen and grind the meat twice.

Line a large tray with parchment paper. Divide the forcemeat into about 450g portions, flatten each portion into a patty about 2.5cm thick, and place them on the prepared tray. You should have 5 patties. Freeze for about 1 hour, until not quite frozen but crunchy at the edges.

Prepare to emulsify the sausage. Refrigerate the bowl and blade of a food processor for at least 30 minutes, or place them in the freezer for at least 15 minutes. When they are well chilled, affix the bowl to the base of the processor and attach the blade. Remove 1 sausage patty from the freezer, break it into 2.5cm chunks, and put them in the food processor. Process for 3 minutes, slowly pouring in 60ml of the beer as the machine is running. Turn off the processor, scrape the emulsified forcemeat into a large bowl, and place the bowl in the fridge. Repeat the process with the remaining sausage patties. It is important that the meat and the machine remain as cold as possible while you work, in order to emulsify the sausage properly. If the food processor begins to feel hot, pause and chill everything again for at least 20 minutes. When all of the forcemeat has been processed, stir to mix well. Cook a small sample of the mixture in a sauté pan and adjust the seasonings if necessary.

Following the instructions on pages 212–214, case the forcemeat in the hog casings and link into sausages about 20cm long, leaving the links connected. Then cut into bunches of 4 links each, and hang each one from an S hook in the fridge overnight.

The following day, fill a large stockpot with the salted poaching water, place over a low heat and heat to 70°C. Place the 4-link bunches in the poaching water. It may be necessary to adjust the heat, as the temperature will drop slightly when you add the links to the pot. Poach the links slowly for 15–20 minutes, until a thermometer inserted into the centre of a sausage registers 60°C. Meanwhile, prepare a large ice-water bath and line a baking tray with a clean kitchen towel.

When the links are ready, transfer them to the ice-water bath for about 20 minutes, until well chilled. Drain the links, clip them apart, and arrange them on the prepared tray. Refrigerate uncovered overnight. The next day, wrap the sausages tightly in cling film and refrigerate for up to 10 days.

Hot Smoking

Hot smoking slowly cooks sausage, imbuing it with the irresistible flavour of wood smoke. Smoking also has the added bonus of preserving sausage by covering the surface with bacteriostatic and mycostatic compounds that work to inhibit the growth of unwanted bacteria and moulds.

Case your forcemeat in loops or link in lengths of four sausages and hang overnight in the fridge to help form a pellicle. The pellicle is the thin skin or glaze formed by drying that seals the surface of the sausage, helping the smoke adhere and keeping the interior moist during smoking.

Prepare your smokehouse following the guidelines listed on page 303. When the smoker reaches a temperature of 75°C to 80°C, hang the sausages in the smoke-filled cabinet. Space them evenly about 5cm apart, and avoid crowding to allow for even smoking. Most sausage will take about an hour to smoke fully. Sausages are cooked when a thermometer inserted into the centre of a sausage registers about 60°C.

If you plan on eating the sausages hot from the smoker, they are now ready. If you plan on storing them to eat later or to eat chilled, prepare a large ice-water bath and line a baking tray with a clean kitchen towel. Remove the sausages from the smoker and place them in the ice-water bath to stop the cooking and to help prevent shrinkage.

When they are thoroughly chilled, remove them from the ice bath. If your sausage is in links, clip them apart. Place the links or loops on the prepared tray and refrigerate, uncovered, overnight. The next day, wrap them tightly in cling film and refrigerate for up to 10 days.

KOLBÁSZ

There are many types of *kolbász,* the Hungarian word for 'sausage'. This smoked version with paprika is traditionally added to *lesco,* a Hungarian vegetable stew with tomatoes and peppers, but it is also spectacular straight from the smoker, served with rye bread, pickled red cabbage, and hot mustard. **MAKES ABOUT 2.5KG OR 5–6 LOOPS**

2 teaspoons black peppercorns

1 cardamom pod

1 tablespoon dried oregano

40g fine sea salt

30g hot Hungarian paprika

1 teaspoon sugar

1 teaspoon curing salt no. 1

1/4 teaspoon ground mace

2 tablespoons minced garlic

**2kg boneless pork picnic, cut into
    2.5cm cubes**

450g pork back fat, cut into 2.5cm cubes

240ml ice water

3m prepared hog casings (see page 210)

Preheat the oven to 170°C (gas 3).

Spread the peppercorns on a baking tray and toast for 3–5 minutes, until fragrant. Let cool completely, then transfer to a spice grinder, add the cardamom pod and oregano, and grind finely. Combine the freshly ground seasonings, sea salt, paprika, sugar, curing salt, mace, and garlic and mix well. Mix the spices evenly with the meat (see page 200), cover, and refrigerate overnight.

Refrigerate the parts of your grinder until ready to use. Remove the meat from the fridge shortly before you are ready to grind. Separate about a quarter of the leanest pieces of pork and set aside. Line a baking tray with parchment paper.

Following the instructions for grinding on page 200, fit the grinder with the smallest screen and grind all the fat and the remaining pork twice. When the second grinding is complete, stop the grinder, switch to the largest screen, and coarsely grind the lean pork once. Combine all of the pork and mix well by hand for 2–3 minutes, until the forcemeat begins to firm up and feel cohesive. Transfer the forcemeat to the prepared baking tray and flatten into a large patty about 2.5cm thick. Freeze for about 1 hour, until not quite frozen but crunchy at the edges.

Place half the partially frozen meat into the bowl of a stand mixer fitted with the paddle attachment and mix on a low speed for 1 minute while slowly incorporating half of the ice water. Once the water has been absorbed, beat on high speed for 1 minute. Scrape the paddled forcemeat into a bowl and set aside. Repeat the process with the remaining meat, incorporating the rest of the ice water. Mix all of the meat together by hand. Cook a small sample of the mixture in a sauté pan and adjust the seasonings if necessary.

Case the forcemeat in hog casings and form into loops (see pages 212–213 and 216). Hang each loop from an S hook in a cool location or in the fridge overnight. The following day, prepare your smoke house following the guidelines listed on page 303. When the smoker reaches a temperature of 80°C, cook the sausages slowly for about 1–1½ hours, until a thermometer inserted into the centre of a loop registers 60°C. If you want to eat the sausages hot from the smoker, they are now ready. If you want to eat them later, prepare a large ice-water bath and line a baking tray with a clean kitchen towel. Place the sausages in the ice-water bath for about 30 minutes.

When they are thoroughly chilled, remove them from the ice bath, place them on the prepared baking tray, and refrigerate, uncovered, overnight. The next day, wrap tightly in cling film and refrigerate for up to 10 days.

HOT LINKS

Spicy beef hot links, smoked slowly over a hardwood fire, are a Texas barbecue speciality. For a true Hill Country experience, smoke a brisket and a few beef ribs alongside these zesty, plump links. **MAKES ABOUT 2.5KG OR 18–20 LINKS**

2 teaspoons coriander seeds

2 teaspoons black peppercorns

3 allspice berries, ground

2 tablespoons ground chilli flakes

1kg boneless lean beef, cut into
    2.5cm cubes

1kg boneless pork butt, cut into
    2.5cm cubes

500g pork back fat, cut into 2.5cm cubes

35g fine sea salt

1 teaspoon curing salt no. 1

2 teaspoons ground cayenne pepper

3 tablespoons minced garlic

250ml ice water

3m prepared hog casings
    (see page 210)

Preheat the oven to 170°C (gas 3).

Spread the coriander seeds on a baking tray and toast for 3–5 minutes, until fragrant. Let cool completely, then transfer to a spice grinder, add the peppercorns, allspice, and chilli flakes, and grind finely.

Place the beef in a non-reactive bowl. Place the pork and pork fat in a second non-reactive bowl. Combine the freshly ground spices, sea salt, curing salt, cayenne, and garlic, then mix a little less than half of the spices with the beef. Mix the remaining spices with the pork and fat. Cover both bowls and refrigerate overnight.

Refrigerate the parts of your grinder until ready to use. Line a baking tray with parchment paper.

Following the instructions for grinding on page 200. fit the grinder with the smallest screen and grind the pork mixture twice. When the second grinding is complete, stop the grinder, switch to the largest screen, and coarsely grind the beef once. Combine the beef and pork mixture and mix well by hand for 2–3 minutes, until the forcemeat begins to firm up and feel cohesive. Transfer to the prepared baking tray and flatten into a large patty about 2.5cm thick. Freeze for 1 hour, until not quite frozen but crunchy at the edges. Meanwhile, chill the bowl and paddle of your stand mixer in the fridge.

Place half the partially frozen meat in the bowl of the stand mixer fitted with the paddle attachment and mix on low speed for 2 minutes while slowly incorporating half of the ice water. Once the water has been absorbed, beat on high speed for 2 minutes. Scrape the paddled forcemeat into a bowl and set aside. Repeat the process with the remaining meat, incorporating the rest of the ice water. Mix all the meat

IN THE CHARCUTERIE

together by hand. Cook a small sample of the mixture in a sauté pan and adjust the seasonings if necessary.

Following the instructions on pages 212–214, stuff the forcemeat in the hog casings and link into sausages about 20cm long, leaving the links connected. Then cut into bunches of 4 links each, and hang each bunch from an S hook in a cool location or in the fridge overnight.

The following day, prepare your smokehouse following the guidelines on page 303. When the smoker reaches a temperature of 80°C), cook the sausages for about 1–1½ hours, until a thermometer inserted into the centre of a link registers 60°C. Hot links enjoyed right out of the smoker are a revelation. But if you cannot eat them all at once, prepare a large ice-water bath and line a baking tray with a clean kitchen towel. Remove the sausages from the smoker and place them in the ice-water bath for about 20 minutes, until well chilled.

When they are thoroughly chilled, remove them from the ice bath, place them on the prepared baking tray, and refrigerate, uncovered, overnight. The next day, wrap tightly in cling film and refrigerate for up to 2 weeks.

Dried Sausage and Salami

There is the science of curing: a complex collaboration of natural processes, an intricate dance done at a microbial level. Then there is the passion for salami: a secret desire to create an underground lair strung with lengths of drying sausages, the air redolent of garlic and a musty perfume. Somewhere between the science and the passion lies the recipe for great salami.

Salami takes its name from the ancient Greek town of Salamis, which perished around 450BCE. But its style of fermented dried sausage lived on and spread from this coastal port across the Mediterranean. Salami is thought to have fuelled the Roman legions, who further spread its popularity throughout the empire, where regional preferences and availability of ingredients gave birth to countless variations.

Good salami is the result of quality ingredients, proper methods, a beneficial environment, a little patience, and a dash of good fortune. It takes the techniques of sausage making and exposes them to the mysteries of the natural world with the hope that in anywhere from two weeks to six months you will possess a thing of beauty.

Note: Although many cultures prepare their own distinctive versions of air-dried sausages, the Italian term *salami* has been used as a catch-all in the English-speaking world. But in Italy, *salame* (*salami* is the Italian plural) has a much more specific meaning than what we generally ascribe to it. In Italy, a *salame* is a large dried cured sausage, and a variety of words (such as *salametto* or *salsiccia secca*) describe smaller dried cured sausage in pork or lamb casing. At the Fatted Calf, we use the word *salami* when referring to the grander category of 'dried cured sausages' – but we also use it in the more specific, Italian sense of the word (that is, to describe larger air-dried sausages). For smaller dried cured sausages, we employ the term *dried sausage*.

Meat and Fat

The meat mix for salami is generally leaner than that for fresh or hot cooked sausage, about 20 per cent fat to 80 per cent lean. Most salami is made from pork or a mix of pork and beef. Pork picnic or leg cuts often contain enough intramuscular fat to make additional back fat unnecessary. Salami can also be made using other meats such as wild boar, venison, or even duck and goose. For optimal results, use meat that has been trimmed of visible fat and supplement with pork back fat to achieve the proper ratio of meat to fat.

Seasoning and Grinding

Because dried sausages are fermented and cured rather than cooked, seasoning them must be done with precision. They need to contain about 3 per cent salt plus additional curing salt. Both ingredients provide seasoning as well as inhibit the growth of unwanted and dangerous microorganisms. Curing salt also gives these sausages their appealing red colour. For better accuracy, salt is measured by weight rather than volume when making all kinds of salami.

Spices provide distinct flavour profiles but also act as antioxidants and stimulate the growth of (good) lactic bacteria. Often a mix of whole and ground spices is used for making salami. Herbs are always used in their dried form and are often ground along with the spices.

Grind the meat for salami just as you would for fresh and cooked sausage, but mix for a bit longer – 3–4 minutes. The salami forcemeat needs to be mixed well in order to distribute ingredients evenly and to eliminate air pockets that can cause spoilage.

BASIC RECIPE FOR SALAMI

2kg lean boneless pork or a mixture of pork and beef
+ 450g pork back fat
+ 70g fine sea salt
+ 1 teaspoon curing salt no. 2

———

2.5kg basic salami

Casing, Linking, Looping, and Tying

In general, the way you case your salami will be determined by the type of casing you use. And the size of the casing you use will in part determine how long you will age your salami.

Small dried sausages in lamb or hog casings can be formed into links just like fresh sausage, but you will need to pack the casings a little more fully before hanging them to dry in lengths of four to eight links. Similarly, you can use hog casings to make loops. Most dried sausage cased in lamb or hog casings will be ready to eat in 2 or 3 weeks.

All beef casings are quite sturdy and make excellent salami that can be hung for many months.

If you are casing in beef middles, sewn beef middles, or beef caps, remove your well-rinsed, soaked, and inverted casing (see page 210) from the soaking water and thread it on to the nozzle. See photos 1–3, opposite.

If you are using regular, open-ended beef middles, pull 2.5cm of the casing over the opening of the nozzle. Tie a 15cm length of twine around the end of the casing. Wind each end of the twine over the top and tie a second knot on the underside to secure the first knot. Create a loop 2.5cm in diameter by knotting the two loose ends of the twine (see photo 4). You will use this loop to hang your finished salami.

Crank the handle slowly to press the sausage mix into the casing to form a 30cm link (see photo 5). Tie a 15cm length of twine into a knot around the casing just before the end of the stuffed casing. This will leave a small amount of the stuffing on the opposite side of the knot (see photo 6). This ensures that the salami is tightly cased and eliminates an air gap at the end. The stuffing on the other side can be pressed away and used to start the next salami. Wind each end of the twine over the top and tie a second knot on the underside to secure it. Create a loop 2.5cm in diameter by knotting the two loose ends of twine. This second loop is used as a fail-safe should the first loop come undone.

Tie a 15-cm length of twine around the casing about 2.5cm from the end of the nozzle to begin your next salami (see photo 7). Wind each end of twine over the top and tie a second knot on the underside to secure it. Then, using scissors, sever the first salami. Begin stuffing the next salami and repeat the process until all the sausage mix is used.

For sewn middles and caps, the process is similar, but you will not need to tie an end knot to begin because one end is already closed. You can expect to get about two large sausages from each piece of sewn middle or cap.

Hanging Salami for Initial Fermentation and for Drying

The raw sausage becomes 'cured' or finished by undergoing a process of fermentation spawned by microorganisms that acidulate the sausage, that is, lower its pH. Fermentation and curing are not exact sciences. Temperature, humidity, and other environmental factors will affect the outcome.

After your salami is cased, you will need to hang it for a period during which the fermentation will begin. Warm and humid conditions tend to encourage the growth of the bacteria necessary for fermentation.

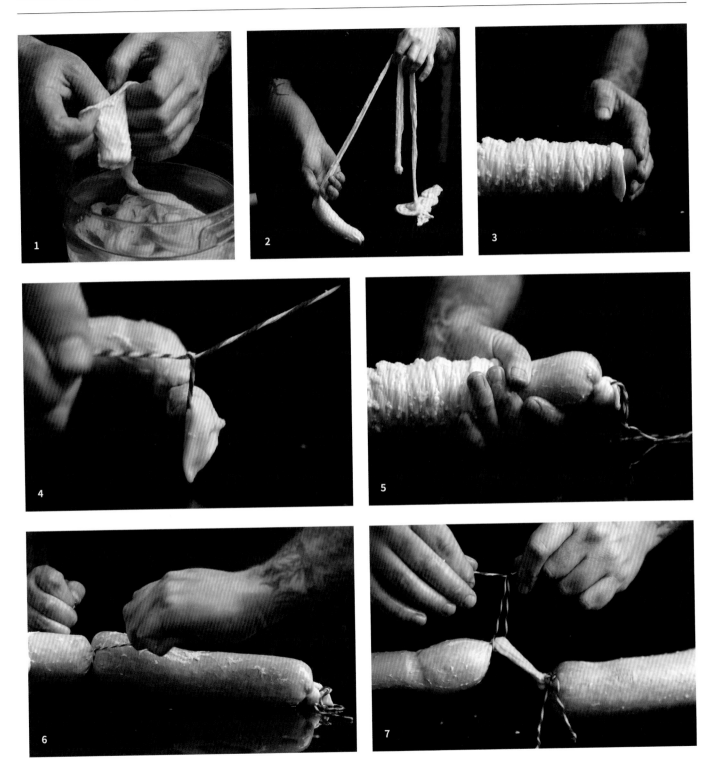

Ideally, a good spot to hang salami has no sunlight and a consistent temperature of 18–24°C. Most dried sausages need to hang for 3–5 days, depending on the size, ingredients, and the conditions of the space (less time in warmer, damper spaces; more time in cooler, drier spaces). If you have a hydrometer, a nifty device that measures the humidity, you are looking to achieve a level of about 70 per cent. If you live in an area with very low levels of humidity, consider using a standard home humidifier to improve conditions. If the conditions are good, a whitish bloom should appear on the outside of the casing in 2 or 3 days. This is the sign that fermentation has started.

If you do not see a bloom, or if the salami starts to appear wrinkled in the first day or two, the environment is most likely too dry. Lightly mist the salami with water once or twice during the initial fermentation period. For dried sausages in beef casings, a daily misting is usually a good idea even if the environmental conditions are good.

Check dried sausages in lamb or hog casings after 3 days. Their colour should be a bright rosy to reddish hue, depending on the spices used, and they should have a pleasing odour. If they are touching one another, an excess of white mould may form, and you will need to gently turn or separate them to improve airflow. Hang them at room temperature, or a bit cooler, to finish drying. For salami in beef casings, check after 3–4 days. Once a healthy white bloom of mould appears, move it into a cooler area (10–16°C) for ageing from 2–6 months.

Allowing and cultivating mould to grow on our food seems to counter what we have come to believe is good and safe, but the moulds on salami work mainly on the outside, creating a barrier that helps to protect them against harmful pathogens during the drying process.

PRESSING LARGE SALAMI

If you want large salami with a very firm texture, pressing after the initial fermentation can help. To press, simply lay the sausages on a tray lined with parchment paper. Lay another piece of parchment over the sausages, then place another tray on top of that. Weight the top tray with something dense and heavy that won't slide off (bricks wrapped in cling film, full olive oil cans, or other rectangular weights are good options). Distribute the weight evenly and press for a week, then hang to cure.

Drying and Ageing

The drying and ageing process is determined by the climate of the curing environment and the size and style of the casing. After fermentation, the sausage has to be dried. This changes the casings from being water permeable to being reasonably airtight, and the process is typically one of some trial and error. Unless you are a commercial salami manufacturer with a highly controlled artificial environment in which you can easily regulate and monitor temperature and humidity, it is unlikely that you will achieve 'perfect' conditions. Focus less on what you cannot achieve and more on what you have to work with. Folks were making salami long before the advent of refrigeration, humidifiers, dehumidifiers, and the like, so it can be done. From drafty uninsulated attics to cool basements, from wine cellars to air-conditioned closets, people have found ways to make it work.

The smaller the casing, the less exact the conditions need to be. Dried sausages in smaller hog or lamb casings dry more quickly and are less subject to environmental fluctuations. If you are making salami for the first time, or attempting to make salami in a new location, make a small batch of dried sausage in lamb or hog casings to test out the conditions of your space. How quickly they dry, whether or not the mould forms or how fast it forms, can clue you in on the relative feasibility.

Larger salami in beef casings require a little more work and care. They are ideally aged around 10°C, but it is more important that you achieve a consistent temperature than an exact one. You can age salami at temperatures as low as 4°C and as high as 16°C, but large fluctuations in temperature can be detrimental to the curing process.

Bacteria: The Good, the Bad, the Commercial

Making salami is not unlike making cheese. During the fermentation process, bacteria digest sugars (naturally occurring from the meat, plus any added sugars) and produce lactic acid. Lactic acid lowers the pH and makes the meat an inhospitable environment for unwanted bacteria. The lactic acid also imparts that tangy flavour generally associated with salami.

Some people use a commercial starter culture or bacteria to help promote the fermentation process and ensure a higher rate of success. Commercial starters produce a highly consistent product but also have a tendency to alter the flavour of the salami, making it very tangy and masking some of the nuances of the flavours of the meat and spices.

You can make salami without a commercial starter. Traditionally made salami depend on wild strains of bacteria usually related to *Lactobacilli plantarum*. These are the natural flora present all around us. Harnessing them can be a little tricky, but we find that the flavour they give salami is far superior to what is possible with commercial starter cultures. Adding a small amount of wine to your salami recipe is another natural way to help to start the fermentation.

You may occasionally find yourself with a bad strain of bacteria. White, grey, green, and even blue

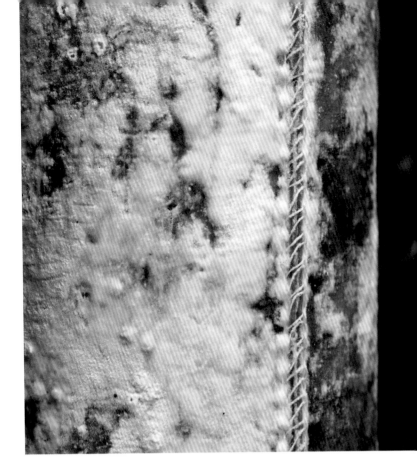

moulds are healthful and beneficial, but if your salami is producing reddish or black mould, don't take any chances. Discard it.

Finishing

Although some indicators can help to determine the readiness of your salami, in truth your salami is done when you decide it is done. Many kinds of salami can be enjoyed at various stages of ripeness, from very young and soft (the style known in Italian as *morbido*, enjoyed in many parts of the country) to very firm and dry. A dry salami will lose about 30 per cent of its original weight and should yield slightly but feel firm when pressed with your fingertips. When you harvest your salami, wipe off any excess mould and peel away the casing before slicing. (Note that you do not need to peel away the casing from dried sausage in hog or lamb casings.) When you slice into a finished salami, it should look brightly coloured and smell irresistible.

CACCIATORINI

Cacciatori are 'hunters' in Italian, and *cacciatorini* are the little dried sausages that hunters can stash easily in their pockets for immediate sustenance during their long treks through the woods. *Cacciatorini* are great fun to make and a perfect starting point if you are just beginning to learn to make salami. **MAKES ABOUT 1.5KG OR 20 (12CM) LINKS**

1 teaspoon aniseeds

1 tablespoon black peppercorns

2.5kg boneless lean pork from shoulder or leg, cut into 2.5cm cubes

60g fine sea salt

1 teaspoon curing salt no. 2

1¹/₂ tablespoons whole chilli flakes

1 tablespoon finely minced garlic

120ml red wine

3m prepared hog casings (see page 210)

Preheat the oven to 170°C (gas 3).

Spread the aniseeds on a baking tray and toast for 3–5 minutes, until fragrant. Let cool completely, then transfer to a spice grinder, add the peppercorns, and grind finely.

Place the meat in a large non-reactive bowl or container. In a small bowl, combine the freshly ground spices, sea salt, curing salt, chilli flakes, and garlic. Mix the spices evenly with the meat (see page 200), cover, and refrigerate overnight.

Refrigerate the parts of your grinder until ready to use. Following the instructions for grinding on page 200, fit the grinder with the smallest screen and grind the meat once. Pour in the wine and mix well by hand for 3–4 minutes, until the meat is very firm and all of the wine is incorporated.

Following the instructions on pages 212–214, stuff the forcemeat tightly in the hog casings and link into sausages 12cm long, leaving the sausages connected. Separate each link from the next by tying off the links with twine, then cut into lengths of 3 links. Hang in a suitable location at or just below room temperature (18-21°C is optimal) for 12–14 days, until firm enough to slice. The finished *cacciatorini* can be stored in a cool, dry spot for several months, to be enjoyed on long walks through the woods and other adventures.

SAUCISSE SEC AUX HERBES DE PROVENCE

These skinny dried sausages seasoned with a profusion of herbs are redolent of a summery meadow. They make a great addition to a picnic basket. **MAKES ABOUT 24 (20CM) LINKS**

2.5kg boneless lean pork shoulder or leg, cut into 2.5cm cubes

1 tablespoon Herbes de Provence (page 15)

1 teaspoon black peppercorns

1 teaspoon white peppercorns

50g fine sea salt

1 teaspoon curing salt no. 2

1 teaspoon fennel pollen or toasted and finely ground fennel seeds (see page 11)

1 tablespoon finely minced garlic

60ml dry white wine

6m prepared lamb casings, 22–24mm in diameter (see page 210)

Place the meat in a large non-reactive bowl or container. In a spice grinder, combine the herbes de Provence and the black and white peppercorns and grind finely. Transfer to a small bowl, add the sea salt, curing salt, fennel, and garlic, and mix well. Mix the spices evenly with the meat (see page 200), cover, and refrigerate overnight.

Refrigerate the parts of your grinder until ready to use. Following the instructions for grinding on page 200, fit the grinder with the smallest screen and grind the meat once. Pour in the wine and mix by hand for about 3–4 minutes, until the meat is very firm and all of the wine is absorbed.

Following the instructions on pages 212–214, stuff the forcemeat tightly in the lamb casings and link into sausages 20cm long, separating each link with a piece of twine. Cut into lengths of 6 links. Hang in a suitable location at, or just below, room temperature (18–21°C is optimal), using 2.5-cm S hooks placed between the second and third, and fourth and fifth, links. Make sure none of the links is touching. The sausages will dry in 1–2 weeks, depending on the ambient temperature. They are ready when they are firm enough to slice.

The finished sausages can be stored in a cool, dry spot for 6 weeks.

PEPPERONI

This spicy, robust salami will make you rethink what you thought you knew about pepperoni. While it certainly can be strewn atop a pizza, or nibbled with olives and cheese, it's also delicious added to a salad of shaved fennel and rocket, or sautéd with thinly sliced cima di rapa. **MAKES ABOUT 3KG OR 10–12 (25CM) SAUSAGES**

1¹/₂ teaspoons fennel seeds

¹/₂ teaspoon aniseeds

1 teaspoon peppercorns

8 allspice berries

4.5kg boneless beef chuck, about 80 per cent lean meat and 20 per cent fat, cut into 2.5cm cubes

150g fine sea salt

2 teaspoons curing salt no. 2

2 tablespoons ground cayenne pepper

3 tablespoons unsmoked Spanish paprika

3 cloves garlic, pounded to a paste in a mortar with ¹/₂ teaspoon fine sea salt

60ml dry red wine

3m prepared beef middles or sewn beef middles (see page 210), rinsed and turned inside out

Preheat the oven to 170°C (gas 3).

Spread the fennel seeds and aniseeds on a baking tray and toast for 3–5 minutes, until fragrant. Let cool completely, then transfer to a spice grinder, add the peppercorns and allspice, and grind finely.

Place the beef in a large non-reactive bowl or container. In a small bowl, combine the freshly ground spices, the sea salt, curing salt, cayenne, paprika, and garlic and mix well. Mix the spices evenly with the meat (see page 200), cover, and refrigerate overnight.

Refrigerate the parts of your grinder until ready to use. Following the instructions for grinding on page 200, fit the grinder with the smallest screen and grind the meat twice. Pour in the wine and mix well by hand for 4–5 minutes, until the meat is very firm and all of the wine is well incorporated.

Following the instructions on page 232, stuff the forcemeat in the beef casings and link into sausages about 25cm long, separating each link as it is cased. Hang in a suitable location with a temperature of 18–21°F for 3–4 days, until the salami has turned a deep red and shows the beginnings of a white bloom on the casing, then move to a cooler location (10°C would be optimal) and hang for 3–4 months, or until firm enough to slice easily with a chef's knife. Once fully cured, these can be stored refrigerated for up to 1 year, but are at their best in the first 3 months.

To serve, wipe away any excess mould with a dry cloth or towel. If you will be using only part of a sausage, using a sharp knife, score the casing around its circumference partway from the end, then peel away the casing from the area to be sliced.

SBRICIOLONA

Sbriciolona, a type of Italian fennel salami, can be loosely translated as 'crumbly thing.' This full-flavoured salami is made with finely ground and coarsely ground pork, which yield a unique texture. Traditionally, it is not pressed and is eaten when it is still quite young and soft, but it is equally good aged longer until firm. **MAKES ABOUT 3KG OR 2 LARGE SALAMI**

4.5kg boneless pork picnic or leg meat cut into 2.5cm cubes

150g fine sea salt

2 teaspoons curing salt no. 2

$1/8$ teaspoon ground mace

2 teaspoons whole chilli flakes

2 teaspoons ground chilli flakes

1 tablespoon fennel pollen

5 cloves garlic, pounded to a paste in a mortar with 1 teaspoon fine sea salt

120ml dry red wine

2 prepared beef caps, turned inside out (see page 210)

Place the pork in a large non-reactive bowl or container. In a small bowl, combine the sea salt, curing salt, mace, whole and ground chilli flakes, fennel pollen, and garlic and mix well. Mix the spices evenly with the meat (see page 200), cover, and refrigerate overnight.

Refrigerate the parts of your grinder until ready to use. Following the instructions for grinding on page 200, fit the grinder with the smallest screen and grind half of the meat once. Stop the grinder, switch to the largest screen, and grind the remaining meat once. Combine both batches of pork, pour in the wine, and mix by hand for about 4 minutes, until you can pick up most of the mix with both hands without it falling apart.

Following the instructions on page 232, case the forcemeat in the beef caps, making sure there are no air holes, then truss tightly with twine as you would a roast (pages 168–9). Hang the sausages in a suitable location at 18–21°C) for 4 days, until they have turned a rosy hue, and ideally have a little bit of a white bloom of mould, then move to a cooler location (10°C would be optimal) and hang for about 4 months, until the desired firmness has been achieved. These salami will hold for up to 10 months but are best within the first 4 months after they're finished curing.

To serve, wipe away any excess mould with a dry cloth. If you will be using only part of a sausage, using a sharp knife, score the casing around its circumference partway from the end, then peel it away from the area to be sliced. Store partially cut *sbriciolona* refrigerated, with the cut end loosely wrapped in parchment paper.

6

PÂTÉS:
POTTED MEATS,
TERRINES & LOAVES

Pâtés, which include potted meats, terrines, and loaves, are quintessential members of the charcutier's repertoire. They sparkle enticingly from shop windows in Paris, provide the delicious makings for a picnic, and summon you to the table to enjoy a leisurely meal. These savoury and addictive meaty goods run the gamut from the haute and the elegant, such as the seductive Duck Liver Mousse (page 271), to the rustic and the everyday, such as the humble American-style Meat Loaf (page 254). They embrace a wide range of techniques as well, from the surprisingly simple salt 'cooked' Foie Gras Torchon (page 275) to the challenging four-day, multiprocessed Head Cheese (page 248). In other words, every curious cook and charcuterie devotee, from novice to master, will find something to pursue in this chapter.

Tools of the Trade

The recipes in this chapter require little special equipment. A grinder is handy for some preparations but not essential for all. For making terrines and loaves, a classic, 1.5-litre rectangular terrine mould made of enamelled cast iron with a tight-fitting lid is sturdy, attractive, ensures even cooking, and will last for many years. But if you don't have a terrine, a bread pan or small baking dish will do. A *tamis*, or drum sieve, is the key to a velvet-textured mousse, but a fine-mesh sieve will work. For potted meats, ramekins, canning jars, ceramic pots, and mason jars are all useful, but one can often be substituted for the other.

Potted Meats

Potted meats are a simple type of pâté made with cooked meats that have been shredded or chopped and bound either with fat or a gelatinous cooking broth and then set in a ceramic pot for serving or keeping. The notion of potted meats seems somewhat antiquated, but they are handy for the busy, modern-day host to serve for hors d'oeuvres. They can be made ahead, then sealed and stored for weeks or months, so that the next time unexpected guests drop in, you can just reach into your larder for the makings of a fabulous canapé or first course.

The recipes that follow are divided into two categories: potted meats bound with fat, and potted meats bound with rich broth.

Potted Meats Bound with Fat

Rillettes and other potted meats bound with fat are delicious preparations that can be made easily if you have confits on hand and are worth the effort even if you don't. Simply season the meat, cook it long and slow in its own fat or in another lipid of your choosing, and then shred and bind it with some of its cooking juices and fat. Pack the mixture into a pot or terrine while it is still warm, and cover it with a layer of fat, which will harden as it cools to form an airtight seal that protects the meat, allowing it to ripen and keep longer.

Nearly any type of meat can be prepared and preserved using this method. Pork, duck, and goose are the most common, but rabbit, chicken, turkey, or other fowl work beautifully, as well. Season the meat a day ahead as you would for making a confit (see page 51) and cook it the following day – again as you would a confit – slow and low and completely submerged in fat. If you are cooking on the hob, your pot should never quite reach a simmer. If using the oven, a temperature around 120°C (gas ½) will do the trick. Continue cooking past the point you would for a confit, until the meat slips easily from the bone, or in the case of boneless meat, until you can pull it apart with almost no effort.

Once the meat is cooked, allow it to rest in the fat for 20–30 minutes, until cool enough to handle. Separate the meat from its cooking liquid and transfer to a large, shallow bowl. Pour the cooking liquid into a tall, clear container to cool. The fat will separate from the *gelée*, the rich meaty juices that accumulate during the cooking process. Ladle the fat off into another container, and set the fat and *gelée* aside separately while you shred or chop the meat.

TASTING FOR SEASONING

When sampling charcuterie before cooking, it is important to taste it at the temperature at which it will be served. Cold food needs to be more highly seasoned, whereas food that will be served hot needs less seasoning.

In most cases, the best texture is achieved if you shred by hand, although there are times when a coarser texture may be desired, in which case you may want to chop the meat with a knife. Avoid using a food processor as it is all too easy to overwork your meat and wind up with an unattractive paste. The meat should be tender and shred easily. Discard any bones, skin, gristle, or cartilage. When the meat is uniformly shredded to your liking, fold in about half the *gelée* and some of the fat by hand. How much fat and liquid you add will depend on the type of meat, its water content, and a host of other factors, including your own preferences.

Working fat-bound meats is a little like working bread dough. The action of your hands helps to make a cohesive meat mass. Incorporating the fat slowly gives the meat time to absorb it and will result in a creamier texture. Use your fingers like the tines of a fork to work in the fat. Try not to mash the meat; the resulting mixture should feel light and loose rather than pasty. When the mixture feels silky and as though it cannot hold any more fat, pause for a bit. If the mixture dries out on the surface and edges after a few minutes, consider adding more fat. If small pools of fat begin to form on the surface, you have reached the point of perfection.

Add freshly chopped herbs such as thyme, oregano, chives, flat-leaf parsley, or rosemary along with wine, brandy, or other flavourings. Now taste for seasoning. Fat-bound potted meats are usually served cool. Cooler temperatures mute the flavours, so you want it to taste highly seasoned while it is still warm. If it needs a little swagger, you can add more of the *gelée*, a pinch more of the seasoning spices, or more salt.

Traditionally, potted meats bound with fat were sealed under a layer of their cooking fat for longer keeping. Although you can eat them right away, sealing and storing allows them to ripen somewhat and develop flavour. Pack or pot the meats into clean glass jars with lids (mason jars are perfect) or a ceramic pot or terrine with a fitted lid, pressing the meat in as you go to eliminate any air pockets. Chill for 1 hour in the fridge, then pour room-temperature cooking fat on top, creating a layer 1cm thick. Cover and refrigerate. After a few hours, check the fat seal. Fat will sometimes continue to seep into the meat as it settles. If any of the meat is poking up or visible through the fat, pour another thin layer of liquid fat on top. Well sealed and refrigerated, potted meats will keep in the fridge for several months.

To serve, remove the jar or pot from the fridge and allow it to warm to room temperature. The top layer of fat can be scraped off and reused to reseal any leftovers, repurposed for cooking, or, in a classically decadent fashion, stirred into the meat and savoured.

VARIATIONS ON POTTED MEATS BOUND WITH FAT

Once you learn the basic method for making potted meats bound with fat, you can try making one of these tasty variations.

Duck Rillettes: Make with whole duck or duck legs simmered in rendered duck fat and seasoned with *quatre épices*, thyme, and dry sherry.

Guineafowl Rillettes: Make with whole guineafowl or legs simmered in rendered duck or chicken fat and seasoned with fresh summer savory, white wine, and roasted peppercorns.

Goose Rillettes with Black Truffle: Simmer goose legs and wing joints in rendered goose or duck fat and season with crushed garlic, Cognac, and grated fresh black truffle. Garnish individual pots with a shaving of black truffle.

Smoked Potted Pork: Use a combination of slab bacon and boneless pork shoulder simmered in lard and finish with a splash of bourbon.

RABBIT RILLETTES

For most potted meats bound with fat, such as those made with pork or poultry, we prefer to cook meat on the bone – but rabbit is full of tiny bones that can be easy to overlook when you are shredding the meat. For this recipe, bone the rabbit and save the carcass for making a flavourful broth (page 44). **MAKES 1.5–2 LITRES**

1 rabbit, 1.4–1.8kg

1 teaspoon peppercorns

3 allspice berries

1 whole clove

1 juniper berry

35g fine sea salt

2 cloves garlic, smashed

1 dried bay leaf

60ml Madeira

2 litres rendered duck fat or lard, or a combination

15g finely chopped fresh sage

Remove the heart, liver, and kidneys from the rabbit and reserve for Marsha's Grilled Rabbit Spiedini with Olives, Almonds and Leaf Salad (page 165). Following the instructions on pages 86–8, bone all of the rabbit except for the front legs and place the meat and front legs in a bowl.

In a spice grinder, combine the peppercorns, allspice, clove, and juniper and grind finely. Transfer to a small bowl, add the salt, garlic, bay, and Madeira, and stir to mix. Add the spice mixture to the rabbit and mix to coat evenly. Cover and refrigerate overnight.

Preheat the oven to 120°C (gas ½). Place the rabbit in a tall, narrow ovenproof pot and cover with the fat. Warm over a low heat on the hob for about 10 minutes, until the fat has just melted, then transfer to the oven to cook slowly for 2–3 hours, stirring gently every 30 minutes. To check if it's cooked, remove one of the front legs and pull some of the meat from the bone. It should yield completely, leaving the bone fairly clean. When it is ready, remove the pot from the oven, pull the meat off the other front leg, discard the bones, and return all the meat to the pot.

Let the meat cool for about 20 minutes then, using a slotted spoon, transfer the rabbit meat to a bowl. Strain the fat through a fine-mesh sieve into a tall, clear container. Let stand for about 20 minutes, until the fat separates from the *gelée*. Ladle off the fat into its own container, and set the *gelée* and fat aside separately.

Shred the meat by hand, then work in the *gelée* and fat, about 120ml at a time, alternating between *gelée* and fat. When the rillettes feels creamy, loose, and cannot absorb any more of the fat (look for small pools of fat on the surface of the meat), test for seasoning. Refrigerate a sample of the rillettes for 15 minutes. Evaluate the taste and texture and adjust the seasonings if necessary. If it seem grainy or dry, add more fat. Fold in the sage.

Pack or pot the rillettes into clean glass jars, ceramic pots, or a terrine, leaving a 1cm head space. Chill for 1 hour in the fridge, then pour some of the room-temperature cooking fat on top, creating a layer 1cm thick. After 1 hour, check the fat seal following the instructions on page 243, then cover and refrigerate.

If the rillettes are well sealed, they will keep well for several months. The flavour will actually improve after the first week. Any extra cooking fat can be strained and refrigerated for several months to be reused for another batch of rillettes, or other low-temperature cooking, such as confits (see page 51).

CICCIOLI

Ciccioli is a traditional Italian preparation concocted from all of the piggy bits, scraps of meat, fat, and skin left over from hog butchering. But there is little consensus as to the method and seasoning used, and there seem to be as many recipes for *ciccioli* as there are picturesque, pork-adoring villages in Italy. This version is a spicy, chunky yet spreadable pâté prepared similarly to rillettes but with a coarser texture and the added benefit of pork crackling. Any leftover spicy lard can be reused several times and is great for cooking. Try using it in place of regular lard for a devilish version of the Chicken-Fried Quail (page 83). You will need to marinate your meat overnight before cooking. **MAKES ABOUT 2 LITRES**

1.5kg boneless skin-on pork shoulder, in a single piece
35g fine sea salt
1½ teaspoons freshly ground pepper
1 tablespoon whole chilli flakes
1 head garlic, cloves separated, peeled, and crushed
3kg lard
2 teaspoons finely ground chilli flakes
60ml dry white wine

Skin the pork shoulder (see page 124, Skinning the Pig) and set aside the meat. Using the tip of a sharp knife and a light touch, score the surface of the skin, making straight lines spaced about 2.5cm apart.

Cut the shoulder meat into 2.5cm cubes and place in a large bowl. Season with the salt, pepper, whole chilli flakes, and garlic, coating the meat evenly. Cover and refrigerate overnight.

The next day, preheat the oven to 200°C (gas 6).

In a large, heavy casserole dish, melt the lard over a low heat. Rub the skin of the shoulder with 1 teaspoon of the melted lard and set the skin, fat side down, on the rack of a roasting tin and place in the oven. Cook for about 30 minutes, until the skin turns golden brown. Turn down the oven temperature to 140°C (gas1) and continue cooking for 1–1½ hours, until crisp and brittle. Remove from the oven, pour any accumulated fat from the roasting tin into the pot of lard, and let the meat cool.

Add the cooled pork shoulder and its marinade to the casserole dish and place over a very low heat. Cook for about 3 hours, making sure the pot never reaches a simmer. The meat is ready when it can be split easily with a fork. Remove the pot from the heat and let rest for 20 minutes.

Using a slotted spoon, transfer the meat to a large bowl. Strain the fat through a fine-mesh sieve into a tall, clear container. Put any solids caught in the sieve, such as spices or bits of meat or garlic, with the pork. Let the fat stand for about 20 minutes, until the fat separates from the *gelée*, then ladle off the fat into its own container and set the *gelée* and fat aside separately.

Pour 120ml each of the fat and *gelée* into the bowl of pork. Coarsely shred the meat by hand while working in the liquid, then mix in the ground chilli flakes and wine. Break the cooked skin into roughly 1cm pieces, either by hand or with a knife, and fold into the shredded pork. Then add another 120ml each of the fat and *gelée*. Refrigerate a sample of the mixture for 15 minutes. Evaluate the taste and texture and adjust the seasonings if necessary. If the mixture seems grainy or dry, add more fat.

Pack or pot the mixture into clean glass jars, ceramic pots, or a terrine, leaving a 1cm space between the top of the meat and the lip of the vessel. Chill for 1 hour in the fridge, then pour some of the room-temperature cooking fat on top, creating a layer 1cm thick. After 1 hour, check the fat seal following the instructions on page 243, then cover and refrigerate.

If the *ciccioli* is well sealed, it will stay good for several months. The flavour will actually improve after the first week. Any extra cooking fat can be strained and refrigerated for several months, to be reused for another batch of *ciccioli* or repurposed for cooking.

Potted Meats Bound with Rich Broth

Potted meats bound with broth are a rather different concoction, more akin to a delicious, rich meaty braise eaten cold. Although the process is similar to potted meats bound with fat, these meats draw their soul power from their flavourful, gelatinous cooking liquid. The seasoned meat is cooked until very tender, then shredded or chopped, bound with broth, packed in a terrine, and chilled to solidify, resulting in a mosaic of meat and seasonings suspended in a flavourful, natural gelatine.

For the best broth-bound potted meats, choose stewing or braising cuts from flavourful, well-developed muscles, preferably on the bone. Nearly any meat can be prepared in this fashion, but those tougher, bonier cuts that contain the most gelatine, such as beef, veal, lamb, or pork shanks, shoulders, heads, and cheeks, will yield the best results. The addition of a split pig trotter or calf foot will give the cooking liquor a tremendous amount of natural gelatine that will act as a binder.

Season your meat ahead as you would for a braise. Brining the meat in advance of cooking is sometimes recommended for particularly dense or bony cuts, such as pork trotters or head. Simmer gently in a flavourful meat broth with aromatics until very tender or the meat slips easily from the bone. Allow to cool to room temperature in the cooking liquid. Drain the meat and transfer to a large, shallow bowl, reserving the liquor separately. Discard any bones and aromatics.

Return the braising liquor to a pan over a hot heat and reduce by about half. Test the potency of the gelatine in the liquid by ladling liquor into a ramekin or shallow bowl to a depth of 1cm, then place in the fridge for about 20 minutes. If it is firm when chilled the liquor has adequate gelatine; if it is still liquid or holds together loosely, continue to reduce the liquor to concentrate the gelatine. Be sure to taste as it reduces as the salt and seasonings will concentrate as the liquid evaporates. If it begins to seem overly seasoned, stop reducing regardless of the results of the gelatine test. It is better to have a loosely bound potted meat than an overly salted one. Allow the finished liquor to cool to room temperature. (Note: Some cooks like to add commercial gelatine as an insurance policy or shortcut. Commercial gelatine, often of dubious origins, is generally unnecessary for potted meats, providing you use a sufficient amount of meat, good broth, and take the time to reduce the cooking liquor.)

When it is cool enough to handle, shred or chop the meat, discarding any bone, skin, or sinew. Add any herbs and/or garnishes. Ladle in small amounts of the braising liquor to moisten and mix gently by hand. You need to add just enough liquid to bind the meat. Refrigerate 1 tablespoon of the mixture for 30 minutes. Evaluate the cohesiveness and the taste and adjust the seasonings if necessary.

Pot your meat in glass jars, pots, or terrines. If you plan on turning out your potted meat (that is, not serving it in the vessel it was stored in), line the terrine or other vessel with cling film, allowing it to overhang the sides by at least 5cm. If you will be serving your potted meat *en terrine* you can skip this step. Either way, ladle a small amount of the cooking liquid into the terrine or other vessel, enough to coat the bottom to a depth of about 1cm. You will need about 120ml. Follow with a layer of roughly half the meat, pressing it down gently. Cover with a little more liquid (again about 120ml), followed by the remainder of the meat. Press gently again, then top with a little more liquid to cover (about 240ml). You will rarely need all of the cooking liquid to set your potted meat and the remainder can be saved for making soups or sauces. Cover the vessel (if you have lined the terrine with cling film, bring the overhang up to cover) and refrigerate overnight to set before serving.

To turn out a terrine, remove it from the fridge and invert it on to a cutting board, then carefully lift off the vessel. Do not bring it to room temperature before you unmould it. Carefully peel away the cling film, then slice the meat. If you will be serving the slices later, transfer them to a baking sheet lined with parchment paper and return to the fridge until serving time. If you are serving the potted meat *en terrine*, just insert a spoon and enjoy.

VARIATIONS

Jambon Persillé: A Parisian classic! Simmer a whole brined ham, 1 onion and a few split pig trotters until quite tender. Shred the meat and season with chopped capers, cornichons, tarragon, and shallots; a splash of white wine vinegar or champagne vinegar; and lots of chopped fresh parsley, *bien sur*!

Pork-cheek Terrine: Brown pork cheeks and braise them in a rich pork broth seasoned with chilli flakes, minced onion, chopped garlic, fresh rosemary, savory, and thyme. Leave the cheeks whole, set them in a rectangular terrine, chill, and slice to serve.

HEAD CHEESE

Admittedly, head cheese is not for everyone, but it does have its devotees, and we see more and more people converted all the time. At the Fatted Calf, we make a batch of head cheese every six weeks, and when it is gone, it is gone until we have amassed enough pork heads to make another batch. Head cheese, though not difficult, is a time-consuming labour of love. This version takes four days from start to finish. On the first day, you make brine, then you brine the heads the following day. This is followed by a day of cooking, chopping, and setting. The terrine chills overnight until finally you make a wish, turn out your terrine and, if the stars have aligned, you have a mosaic of head meat of your own making that slices beautifully. Serve with baguette, cornichons, olives, and mustard, or offer as part of a salad course with spicy cress dressed with a pungent vinaigrette. **MAKES ABOUT 3 LITRES OR TWO 1.5KG TERRINES**

HEAD BRINE
1 tablespoon black peppercorns
1 tablespoon fennel seeds
1 juniper berry
1 dried bay leaf
450g fine sea salt
340g sugar
6.5 litres boiling water
2 teaspoons curing salt no. 1

MEAT
1 skin-on pig head, split or quartered and brain removed
3 pig's trotters, split
1kg bone-in, skin-on pork shoulder, in a single piece

GARNISH
1 tablespoon fennel pollen
2 teaspoons chilli flakes
1 teaspoon freshly ground black pepper
1/8 teaspoon ground mace
Grated zest and juice of 1 lemon
60ml dry white wine
15g chopped fresh flat-leaf parsley

Day 1 Make the brine. Place the peppercorns, fennel seeds, juniper berry and bay leaf on a square of cheesecloth, bring the corners together and tie securely with twine to make a sachet (or use a muslin bag). Measure the sea salt and sugar into a 20-litre non-reactive bucket and toss in the sachet. Pour in the boiling water and stir to dissolve. Leave to cool overnight.

Day 2 Stir the curing salt into the brine. Add the meat and top with a plate to keep it submerged. Refrigerate for 24 hours.

Day 3 Drain the meat, discarding the brine and sachet. Rinse lightly under cool running water and pat dry, then put the meat into a tall, narrow pan and cover with water. Bring to the boil on the hob over a high heat, then reduce the heat to a lively simmer and skim any impurities that rise to the surface. Cook, uncovered, for 3 hours, then check the head for tenderness. The meat should begin to pull away from the bone and the ears should be softened.

Carefully transfer the meat to a tray to cool. Line a colander with cheesecloth (or use a chinois or other fine-mesh sieve) and strain the broth into a stockpot. Cook on the hob over a medium heat until reduced by half, then ladle into a ramekin to a depth of 1cm and refrigerate for about 20 minutes. If it is firm and gelatinous when chilled, it is ready. If it is still a touch runny, reduce further and test again. Adjust seasoning if necessary.

Pull the meat from the bones of the head and trotters. Discard the bones. Slice the ears and trotter skin into 5mm strips. Tear the cheek meat by hand into pieces of roughly the same size, and cut all the remaining head meat and the pork shoulder into 1cm cubes. Place all the meat in a large bowl and add the fennel pollen, chilli flakes, pepper, mace, lemon zest and juice, wine, and parsley. Mix gingerly to incorporate the garnish.

Following the instructions on pages 246–47, pot the meat in mason jars, or 2 terrines, lining the terrines with cling film if you plan on turning out the potted meat to serve. Refrigerate overnight.

Day 4 Serve as directed.

OXTAIL TERRINE

The bony, crosscut *queue de boeuf* (French for 'oxtail') is teeming with earthy flavour and gelatine, which is the key to its success. Top slices with a strong mustard and caper sauce and serve alongside a handful of spicy cress. **MAKES 1.5–2 LITRES OR ONE 3-POUND 1.5KG TERRINE**

2.3kg oxtails, cut crosswise into pieces
   5cm thick
Fine sea salt
2 tablespoons lard
480ml dry red wine
480g whole tinned tomatoes with their juice
1 pig's trotter, split
1 onion, quartered
2 carrots, peeled and sliced on the diagonal
   2.5cm thick
5 cloves garlic, peeled but left whole
1 dried bay leaf
¹/₂ teaspoon black peppercorns
1 teaspoon piment d'Espelette
30g chopped fresh flat-leaf parsley

Rinse the oxtails under cool running water and pat dry. Place in a large bowl and season liberally with salt. Cover and refrigerate overnight.

The next day, melt the lard in a large, heavy sauté pan over a medium-high heat. Add the oxtails in a single, uncrowded layer and brown well on all sides. Using a slotted spoon, transfer the oxtails to a large casserole. Pour the wine into the sauté pan and deglaze over a medium heat, stirring with a wooden spoon to loosen the fond from the bottom of the pan. Pour the wine over the oxtails, then add the tomatoes, pig's foot, onion, carrots, garlic, bay leaf, and peppercorns. Add water just to cover the oxtails and 1 teaspoon salt and bring to a boil. Lower the heat to a lazy simmer and skim off any particulates that rise to the surface. Simmer, uncovered, for 3 hours.

When the meat is fork-tender, transfer it to a platter. Line a colander with cheesecloth (or use a chinois or other fine-mesh sieve) and strain the broth into a stockpot. Place the pot over a medium heat and cook until reduced by about a quarter. Ladle the broth into a ramekin to a depth of 1cm and refrigerate for 20 minutes. If it is firm and gelatinous when chilled, it is ready. If it is still a touch runny, reduce further and test again. Taste for seasoning and adjust if necessary.

Pull the meat off of the bones, and discard the skin and bones. Toss with the *piment d'Espelette* and parsley.

Following the instructions on pages 246–47, pot the meat in mason jars, or a terrine, lining the terrine with cling film if you plan on turning out the potted meat to serve, then refrigerate and serve as directed.

Baked Terrines and Loaves

A terrine is a type of pâté or forcemeat that is baked in a mould, also called a terrine. A loaf is a forcemeat that is mounded and shaped by hand and baked uncovered to form a crust. Both are made with seasoned meats that are ground and mixed with a binder to help retain their shape, and both are usually garnished with herbs, vegetables, or other raw or cooked meats before they are baked in a low oven. The only difference is the dish or its absence.

Terrines and loaves can be made from almost any meat or poultry but generally contain some pork to help improve the texture and fat content. Ideally, a loaf or terrine contains about 30 per cent fat. Many traditional terrines and loaves contain organ meats such as liver or gizzard, in addition to meat and fat, but it is a common misconception that they *always* contain these or are mainly comprised of them. If you don't care for organ meats, you need not add them. But if you can obtain good-quality fresh organ meats, we urge you to try cooking with them, as they add a subtle depth to the flavour of many preparations.

Marinate the meat a day ahead as you would for most sausages. Prior to grinding, prepare any garnishes to add after the meat is ground. Confit of duck gizzard, olives, brandied fruits, poached tongues, diced liver, braised shallots, chopped herbs, cooked mushrooms, strips of lean meat and the like all make attractive and (delicious) garnishes.

Grind the meat as you would for sausage. Rustic country-style terrines and loaves are usually ground only once, but for finer, lighter-textured terrines you can grind the meat twice. Then fold in your panade or other binder as well as any garnish to be incorporated. In some cases, the garnish can be artfully arranged to create a mosaic in each slice. If you're doing this, reserve the garnishes until the terrine is assembled.

If you are making a meat loaf or rustic pâté loaf that you will shape by hand, line a baking sheet with parchment paper, mix the meat well, transfer it to the prepared baking sheet, and mould into the desired shape. Smooth the edges with your hands to create a seamless, uniform surface to prevent cracking during baking. Refrigerate the loaf for at least 30 minutes before baking to help prevent fat loss during cooking. When chilled, remove the parchment-lined pan from the fridge and transfer it directly to a preheated 150°C (gas 2) oven. Bake until a thermometer inserted into the centre of the loaf registers 60°C. Loaves can be served warm, at room temperature or chilled. Always allow the loaf to rest at room temperature for at least 10 minutes before slicing.

If the pâté is to be cooked *en terrine*, you will need to line the vessel (see box on page 252) before you add the forcemeat. This allows for easy unmoulding once the terrine is cooked. Add your meat to the lined terrine in increments, pressing firmly after each addition. If you are including a garnish, place it between layers of meat or down the centre of the terrine. Pack the forcemeat to the lip of the vessel and fold over the excess lining to cover the meat. If you have any exposed meat, cut an extra piece of lining or a little back fat for a patch job. If you wind up with too much lining, trim it with scissors to minimise overlap. A little overlap is fine as the lining will recede slightly during cooking. Cover the terrine.

Baking and Unmoulding

Baking a terrine in a water bath helps to ensure even cooking. Fold a kitchen towel in half and place it in the bottom of a roasting tin to act as a buffer. Place the loaded terrine on top. Pour simmering water into the pan to reach a third of the way up the sides of the terrine. Carefully place the roasting tin in a preheated 150°C (gas 2) oven. If your terrine has a small hole in its lid, make sure the side with the hole is placed closest to the oven door to make it easier to take the temperature of the terrine during cooking. Cook to an internal temperature of 60°C. Gently lift the terrine from the water bath and let it cool to room temperature before refrigerating.

Pressing is a technique that is sometimes used to compress the terrine so that it can be sliced thinly without crumbling, though it is usually not necessary if the terrine is packed carefully prior to baking. It is preferable to avoid pressing, however, because pressing a terrine leaches out the delicious cooking juices that help to keep it moist. If you need to press a terrine, you must use a weight that is fitted to your terrine mould. A block of wood cut to size and wrapped in cling film works well. Place the block directly on top of the still-warm terrine and then place the lid on top of the block.

To unmould your chilled terrine, uncover and place it in a hot-water bath for about 30 seconds to gently melt the exterior fat and loosen the sides and bottom from the vessel. Remove from the water bath and invert on to a cutting board. If the terrine does not slip easily from the mould, turn it back upright, loosen it gently on either end using a rubber spatula, and then invert again.

Slice as much of the terrine as you plan on serving. The terrine lining, whether it is caul fat, belly, back fat, or skin, is completely edible.

If you don't plan on serving the terrine right away, you can store it refrigerated in its mould for about a week. If you plan to keep your terrine longer, you can cap it with duck or pork fat, much like a confit (see instructions on page 51). Before capping it, remove any *gelée* that has formed in the bottom or on the sides of the terrine.

FOUR WAYS TO LINE A TERRINE

Caul fat: Versatile caul fat (see page 213) is the simplest way to line a terrine and can be cut to fit any size mould. It is thin and transparent, providing a peek at the meat within, yet sturdy. Lay a large piece in your terrine, allowing it to overhang the sides by about 8cm.

Back fat: Strips of thinly sliced back fat make an impressive-looking all-white lining for a terrine. Slice it on a meat slicer, then layer inside the terrine, slightly overlapping each slice.

Belly: Pork belly or thinly sliced bacon make a flavourful, streaky lining for a terrine. When using bacon, keep in mind that it will add to the seasoning of the terrine. Be sure to overlap the slices, as there will be a small amount of shrinkage during cooking.

Duck skin: If you are using a whole duck to make the forcemeat, you can use the skin to line the terrine. You will need to follow the instructions for Whole Boned Bird (pages 74–6), then separate the meat and skin without tearing the skin. With a knife, scrape away any excess fat or glands that are attached to it. Lay the sheet of skin in the terrine, pressing it into the corners.

From front: belly, caul fat, back fat, duck skin ▶

MEAT LOAF

This classic American pâté is glazed with tangy cocktail sauce (which can be made, along with the panade, a day ahead). A thick slice over creamy mashed potatoes is a thing of beauty. Any leftovers make excellent sandwiches with whole-grain mustard and Pickled Red Onion Rings (page 313).

MAKES ONE 1.6KG LOAF; SERVES 6–8

FORCEMEAT

³/₄ teaspoon black peppercorns

2 allspice berries

2 teaspoons fennel seeds

1 teaspoon chilli flakes

1 whole clove

600g boneless lean beef from eye of round or sirloin, cut into 2.5cm cubes

450g boneless pork picnic, in 2.5cm cubes

115g pork back fat, in 2.5cm cubes

1 tablespoon fine sea salt

2 teaspoons chopped garlic

115g bacon, chopped

PANADE

300g diced onion

2 tablespoons unsalted butter

¹/₄ teaspoon fine sea salt

20g fresh breadcrumbs

1 egg, lightly beaten

120g ketchup

1¹/₂ teaspoons Tabasco sauce

1¹/₄ teaspoons Worcestershire sauce

15g chopped fresh flat-leaf parsley

2 tablespoons chopped fresh oregano

1 tablespoon chopped fresh sage

1 tablespoon chopped fresh thyme

COCKTAIL SAUCE

60g ketchup

¹/₄ teaspoon fine sea salt

¹/₄ teaspoon freshly ground black pepper

1¹/₂ teaspoons grated horseradish

¹/₂ teaspoon Worcestershire sauce

¹/₂ teaspoon Tabasco sauce

Preheat the oven to 170°C (gas 3).

To make the forcemeat, toast the peppercorns and allspice on a baking sheet for 3–5 minutes, until fragrant, then cool completely. Repeat with the fennel seeds. Reserve half the fennel seeds and half the chilli flakes. In a spice grinder, finely grind the remaining chilli flakes, fennel seeds, toasted peppercorns, allspice and the clove.

Place all the meat except the bacon in a large non-reactive bowl. In a small bowl, combine the freshly ground spices, salt and garlic. Mix evenly with the meat, cover, and refrigerate overnight. Reserve the bacon in the fridge.

To make the panade, combine the onions and butter in a sauté pan over a low heat. Add the salt and sweat slowly for 20 minutes, until translucent. Transfer to a bowl and cool to room temperature. Add the breadcrumbs, egg, ketchup, Tabasco, Worcestershire sauce, parsley, oregano, sage and thyme to the onion and mix well.

To make the cocktail sauce, combine all the ingredients in a small bowl and mix well.

Preheat the oven to 150°C (gas 2).

Line a baking sheet with parchment paper. Following the instructions for grinding on page 200, fit the grinder with the largest plate and grind the bacon once. Switch to the medium plate and grind the beef and pork once. Mix the ground bacon into the ground beef and

pork. Pour the panade over the meat and mix by hand for 2–3 minutes, until the forcemeat comes together in a cohesive mass. Cook a small sample of the mixture in a sauté pan and adjust the seasonings if necessary.

Transfer the forcemeat to the prepared baking sheet. Using your hands, mould into a long, even loaf about 6cm tall and 20cm wide. Smooth over any cracks so that the surface appears relatively uniform. If the loaf seems at all crumbly, return it to the bowl to knead further and then reshape. Small cracks will become bigger during cooking, making the loaf unattractive and hard to slice.

Using the edge of a spatula or the back of a chef's knife, score the top of the loaf in a diamond pattern with cuts about 3mm deep.

Bake for 15 minutes, then rotate the pan 180 degrees to ensure even cooking. Check the temperature after another 10 minutes. When a thermometer inserted into the centre of the loaf registers 38°C, remove from the oven and spread the cocktail sauce evenly over the top and sides. Return to the oven and continue baking for 10–15 minutes, until the loaf reaches an internal temperature of 60°C.

Let the loaf rest for at least 10 minutes before slicing. Wrapped tightly and refrigerated, the loaf will keep for 3–4 days.

FORAGER'S TERRINE

Foraging for wild mushrooms is one of my favourite pastimes. The duxelles used to flavour and garnish this terrine can be made with whatever wild mushrooms are available. While fresh wild mushrooms are preferred, dried porcini or morels work well, too. **MAKES ONE 1.4KG TERRINE**

FORCEMEAT

5 shallots, finely minced

2 tablespoons unsalted butter

170g pork liver, trimmed, into 2.5cm cubes

480ml whole milk

800g boneless pork shoulder, 2.5cm cubes

400g pork back fat, cut into 2.5cm cubes

$^1/_2$ teaspoon black peppercorns

$^1/_4$ teaspoon yellow mustard seeds

1 dried bay leaf

2 allspice berries

25g fine sea salt

$^1/_4$ teaspoon ground ginger

$^1/_4$ teaspoon piment d'Espelette or ground cayenne pepper

$^1/_8$ teaspoon freshly grated nutmeg

PANADE

120ml meat broth, any kind (see Basic Rich Broth, page 44)

120ml heavy cream

30g fresh breadcrumbs

GARNISH

160g Wild Mushroom Duxelles (see page 21)

2 tablespoons Madeira

140g finely diced bacon

2 tablespoons chopped fresh thyme, plus 3 sprigs

2 tablespoons chopped fresh flat-leaf parsley

2 tablespoons chopped fresh sage, plus 3 whole leaves

1 piece caul fat or sliced pork fat back, about 20 x 50cm

To make the forcemeat, combine the shallots and butter in a sauté pan over a low heat and sweat slowly for about 20 minutes, until tender and translucent. Remove from the heat and leave to cool to room temperature.

In a bowl, combine the liver and the milk, immersing the liver fully. Cover and refrigerate.

Place the pork and back fat in a large nonreactive bowl. In a spice grinder, combine the peppercorns, mustard seeds, bay, and allspice and grind finely. Transfer to a small bowl, add the salt, ginger, *piment d'Espelette*, and nutmeg. Mix the shallots and spices with the meat, cover, and refrigerate overnight.

To make the panade, in a small bowl combine the broth, cream, and breadcrumbs and mix well. Set aside.

Drain the pork liver, discarding the milk, and add to the marinated meat along with 40g of the mushroom duxelles. Following the instructions for grinding on page 200, fit the grinder with the smallest plate and grind finely twice. Fold the panade and Madeira into the ground meat until fully incorporated, then add the remaining duxelles, bacon, and the chopped thyme, parsley, and sage. Cook a small sample of the mixture in a sauté pan, then chill the sample, taste for seasonings, and adjust if necessary.

Preheat the oven to 170°C (gas 3).

Bring a kettle filled with water to a simmer. Line a terrine with the caul fat. Arrange the sage leaves and thyme sprigs along the centre of the terrine and then fill with the forcemeat. Pack the forcemeat well, tamping down after each addition, then tapping the terrine against the work surface to release any air pockets. The forcemeat should be level with the lip of the terrine. Fold the excess caul fat over the top; trim any overlap and cover the terrine.

Fold a kitchen towel in half and place it in the bottom of a roasting tin. Set the loaded terrine on the towel. Pour the simmering water into the roasting tin to reach a third of the way up the sides of the terrine. Carefully place in the oven and bake for about 1 hour, until a thermometer inserted into the centre of the terrine registers 60°C. Remove the roasting tin from the oven and gently lift the terrine from the water bath. Leave to cool to room temperature, then refrigerate overnight.

To unmould the terrine, uncover and place the terrine in a hot-water bath for about 30 seconds to melt the exterior fat gently and loosen the bottom and sides from the mould. Remove from the water bath, invert on to a cutting board and lift off the mould. Slice as much as you plan to serve, then wrap the remainder well and store in the fridge for up to 1 week.

SPICED LAMB TERRINE

Lamb aficionados delight in this boldly spiced terrine with chunks of poached tongue and whole coriander seeds. Serve with grilled flatbread and quince chutney. **MAKES ONE 1.4KG TERRINE**

FORCEMEAT

1½ teaspoons black peppercorns

2 teaspoons cumin seeds

2 allspice berries

2 dried bay leaves

1kg boneless lean lamb shank, 2.5cm cubes

450g pork back fat, cut into 2.5cm cubes

40g fine sea salt

½ teaspoon curing salt no. 1

2 tablespoons Aleppo pepper flakes (pul biber)

1 tablespoon finely chopped garlic

BRINED TONGUES

1 tablespoon black peppercorns

1 tablespoon yellow mustard seeds

3 allspice berries

2 bay leaves

270g fine sea salt

100g sugar

4 litres boiling water

6 lamb tongues or 4 pork tongues

TONGUE-POACHING LIQUID

2 litres lamb or other meat broth (see Basic Rich Broth, page 44)

240ml dry white wine

3 cloves garlic, smashed

1 star anise, toasted (see page 11)

2 dried cayenne peppers

1 x 5x2.5cm piece orange peel

1 tablespoon fine sea salt

PANADE

60ml strained tongue poaching liquid

120ml double cream

15g fresh breadcrumbs

GARNISH

Reserved diced tongues

2 tablespoons coriander seeds, toasted (see page 11) and lightly crushed

2 tablespoons finely chopped fresh flat-leaf parsley

2 tablespoons finely chopped fresh oregano

1 tablespoon Aleppo pepper flakes (pul biber)

1kg pork back fat, sliced into thin 2.5cm wide strips, 20–25cm long, or 1 piece caul fat, about 20 x 50cm

Preheat the oven to 170°C (gas 3).

To make the forcemeat, spread the peppercorns, cumin, and allspice on a baking sheet and toast for 3–5 minutes, until fragrant. Let cool completely, then transfer to a spice grinder, add the bay, and grind finely.

Place the lamb and fat in a large non-reactive bowl. In a small bowl, combine the freshly ground spices, sea salt, curing salt, Aleppo pepper flakes (*pul biber*) and garlic and mix well. Mix the spice kit with the meat, cover, and refrigerate overnight.

To make the brine for the tongues, place the peppercorns, mustard seeds, allspice and bay on a square of cheesecloth, bring the corners together and tie securely with twine to make a sachet (or use a muslin bag). Measure the salt and sugar into a large container and toss in the sachet. Pour in the boiling water and stir to dissolve the salt and sugar. Let cool to room temperature, then add the tongues and top with a plate or other weight to keep them submerged and refrigerate. If using lamb tongues, leave them to brine for 12 hours. For pork tongues, leave them to brine for 24 hours.

CONTINUED

To make the poaching liquid, in a large pan combine the broth, wine, garlic, star anise, peppers, orange peel, and salt and bring to a simmer over a low heat. Cook for about 30 minutes to allow the flavours to meld.

Remove the tongues from the brine and discard the brine. Rinse the tongues briefly under cold running water, then add to the simmering poaching liquid. Cook until tender, about 40 minutes for lamb tongues and 1¼ hours for pork tongues.

Using a slotted spoon, transfer the tongues to a plate to cool. Strain the poaching liquid, reserving enough for the panade. When the tongues have cooled, cut them into 6mm cubes.

To make the panade, in a bowl, combine the poaching liquid, cream and crumbs and mix well.

Following the instructions for grinding on page 200, fit the grinder with the smallest plate and grind the seasoned lamb and fat once. Fold in the panade and grind again. Add the cubed tongues, coriander, parsley and oregano and mix well by hand for 2–3 minutes, until the forcemeat pulls together. Cook a small sample of the mixture in a sauté pan, then chill the sample, taste for seasonings, and adjust if necessary.

Preheat the oven to 170°C (gas 3). Bring a kettle filled with water to a simmer. Sprinkle the *pul biber* evenly over the back fat (or caul fat, if using), then line a terrine with the fat so that the pepper-covered side is pressed against the terrine. Fill the terrine with the forcemeat. Pack the forcemeat well, tamping down after each addition and then tapping the terrine against the work surface to release any air pockets. Fold the excess back fat over the top; trim any overlap and cover the terrine.

Fold a kitchen towel in half and place it in the bottom of a roasting tin large enough to accommodate the terrine. Set the loaded terrine on the towel. Pour the simmering water into the roasting tin to reach a third of the way up the sides of the terrine. Carefully place in the oven and bake for about 1 hour, until a thermometer inserted into the centre of the terrine registers 60°C. Remove the roasting tin from the oven and gently lift the terrine out of the water bath. Leave to cool to room temperature, then refrigerate overnight.

To unmould the terrine, uncover and place in a hot-water bath for about 30 seconds to melt the exterior fat gently and loosen the bottom and sides from the mould. Remove from the water bath, invert on to a cutting board, and lift off the mould.

Slice as much as you plan to serve, then wrap the remainder well and store in the fridge for up to 1 week.

DUCK TERRINE WITH BRANDIED PRUNES

At the Fatted Calf we preserve both fresh and dried fruits in brandy. They make excellent accompaniments to roasted and cured meats and are delectable embellishments for savoury terrines. Brandied prunes, laid down the centre of this duck terrine, are as visually dramatic as they are delicious. Serve whole slices of this terrine with its elegant prune garnish alongside a tender frisée salad or atop warm slices of brioche. **MAKES ONE 1.4KG TERRINE**

FORCEMEAT

700g boneless, skinless duck meat (see pages 72–3), cut into 2.5cm cubes

340g boneless pork shoulder, cut into 2.5cm cubes

225g pork back fat, cut into 2.5cm cubes

1 teaspoon black peppercorns

1/2 teaspoon white peppercorns

2 allspice berries

1 whole clove

1 small dried bay leaf

25g fine sea salt

1/4 teaspoon ground ginger

1/2 teaspoon piment d'Espelette

1/4 teaspoon curing salt no. 1

PANADE

60ml duck broth (see Basic Rich Broth, page 44)

60ml brandied prunes liquor

120ml heavy cream

15g fresh breadcrumbs

GARNISH

2 tablespoons chopped fresh thyme

2 tablespoons chopped fresh flat-leaf parsley

12 brandied prunes (see Dried Fruit in Brandy, page 18)

1 piece caul fat, about 20 x 50cm, or 1 duck skin (see page 252, Four Ways to Line a Terrine)

To make the forcemeat, place the duck, pork, and fat in a large non-reactive bowl.

Combine the black and white peppercorns, allspice, clove, and bay in a spice grinder and grind finely. Transfer to a small bowl, add the salt, ginger, *piment d'Espelette*, and curing salt, and mix well. Mix the spices with the meat, cover, and refrigerate overnight.

To make the panade, in a small bowl, combine the broth, liquor, cream, and breadcrumbs and mix well.

Following the instructions for grinding on page 200, fit the grinder with the smallest plate and grind the meat once. Fold in the panade and grind again. Add the thyme and parsley and mix by hand for 2–3 minutes, until the forcemeat pulls together. Cook a small sample of the mixture in a sauté pan, then chill the sample, taste for seasonings, and adjust if necessary.

Preheat the oven to 170°C (gas 3). Bring a kettle filled with water to a simmer. Line a terrine with the caul fat or duck skin. Pack half the forcemeat into the bottom of the terrine. Lay the prunes in a line down the centre, making sure there are no gaps between them. Pack the other half of the forcemeat on top of the prunes. The forcemeat should be level with the lip of the terrine. Fold the excess caul fat or duck skin over the top; trim any overlap and cover the terrine.

Fold a kitchen towel in half and place it in the bottom of a roasting tin large enough to accommodate the terrine. Set the terrine on the towel. Pour the simmering water into the roasting tin to reach one-third of the way up the sides. Carefully place in the oven and bake for about 1 hour, until a thermometer inserted into the centre of the terrine registers 60°C. Remove the roasting tin from the oven and gently lift the terrine out of the water bath. Let cool to room temperature, then refrigerate overnight.

To unmould the terrine, uncover and place it in a hot-water bath for about 30 seconds to gently melt the exterior fat and loosen the bottom and sides from the mould. Remove from the water bath, invert on to a cutting board, and lift off the mould. Slice as much as you plan on serving, then wrap the remainder well and store in the fridge for up to 1 week.

Galantines

Nothing harks back to the good old days of traditional French haute cuisine quite like a galantine. A bird is boned and stuffed with a forcemeat adorned with slivers of fresh black truffle or studded with nuts or fruit. The whole thing is rolled and tied, then poached and chilled in its cooking liquid, sliced and served cold alongside a little salad or as part of an *assiette de charcuterie* at the start of a meal. *Très classique!*

The word *galantine* is thought to come from the old French word for chicken, *geline*; hence, galantines are primarily made from poultry. Chicken, guineafowl, duck, pheasant and other game birds can all be used to make elegant and impressive galantines. To begin, all of the bones are carefully removed without making any tears in the skin (see Whole Boned Bird, pages 74–7). Next, a flavourful broth is made from the bones that will be used to cook the galantine later. For some galantines, you will leave some or all of the meat attached to the skin; for others, the meat will become part of the stuffing, or forcemeat. Any meat separated from the skin is usually combined with pork or other meats and marinated overnight as you would for sausage or a terrine. The next day, the meat is ground to produce a forcemeat. The livers and gizzards can be ground along with the meat or saved and used as a garnish. The forcemeat is stuffed into the skin, the whole package is shaped into a cylinder and tied or rolled in cheesecloth, and then the galantine is submerged in liquid and slowly poached until it reaches an internal temperature of about 65°C. Finally, the galantine is chilled in its poaching liquid to help it to absorb more flavour from the liquid and to keep it moist and succulent.

To serve the galantine, it is removed from its poaching liquid. If you like, you can reduce the liquid and use it to glaze the galantine. Slice the galantine into rounds and serve.

VEAL AND CHICKEN GALANTINE

This galantine, studded with green olives and seasoned with fresh herbs, makes an impressive first course or luncheon dish. Although its preparation is labour-intensive, it will keep for up to 4 days in the fridge, so feel free to cook it in advance of serving. **MAKES ONE 1.5KG GALANTINE**

FORCEMEAT

1 chicken, 2–2.5kg

340g pork back fat

225g lean veal meat from leg or shoulder

115g pancetta, home-made (see page 295) or shop-bought

1 teaspoon yellow mustard seeds, toasted (see page 11)

³/₄ teaspoon black peppercorns

¹/₄ teaspoon white peppercorns

25g fine sea salt

¹/₄ teaspoon curing salt no. 1

¹/₈ teaspoon ground mace

¹/₂ teaspoon ground cayenne pepper

POACHING LIQUID

Bones from the chicken, plus 1kg additional chicken bones and feet

120ml brandy

1 dried bay leaf

1 tablespoon fine sea salt

PANADE

60ml dry white wine

60ml heavy cream

15g fresh breadcrumbs

GARNISH

70g pitted and sliced picholine or Castelvetrano olives

15g chopped fresh flat-leaf parsley

2 tablespoons chopped fresh thyme

Follow the instructions for Whole Boned Bird on pages 74–6. Once the meat is completely off the bone, trim any glands or blood vessels from the meat. Carefully remove the meat from the skin, again making sure to avoid puncturing holes in the skin. Wrap the skin in cling film and refrigerate.

Weigh 700g of the chicken meat for the galantine and save any extra for another use. Cut the chicken, fat, veal, and pancetta into 2.5cm cubes and place in a large non-reactive bowl.

In a spice grinder, combine the mustard seeds and the black and white peppercorns and grind finely. Transfer to a small bowl, add the sea salt, curing salt, mace, and cayenne pepper and mix well. Mix with the meat, cover, and refrigerate overnight.

To make the poaching liquid, chop the bones and make a simple bone broth (see Basic Rich Broth, page 44). Strain and refrigerate overnight, uncovered. The following day, remove any solidified fat from the surface, then bring the broth to a simmer in a narrow, tall pan. Measure the liquid and reduce if necessary to yield 3 litres. Add the brandy, bay, and salt and taste for seasonings. It should be

CONTINUED

VEAL AND CHICKEN GALANTINE, continued

very flavourful and well seasoned. Set aside until you are ready to poach the galantine.

To make the panade, in a small bowl, combine the wine, cream, and breadcrumbs and mix well. Assemble the garnish ingredients.

Following the instructions for grinding on page 200, fit the grinder with the smallest plate and grind the meat once. Fold in the panade and the garnish and mix by hand for 2–3 minutes, until the forcemeat holds together and the garnish is evenly distributed. Cook a small sample of the mixture in a sauté pan, chill the sample, taste for seasonings, and adjust if necessary.

Spread the skin out on a cutting board, with the longest edge closest to you. Lay the forcemeat lengthwise down the middle and mould into a cylinder. Fold the longest edge of the skin over the forcemeat and roll the cylinder away from you to cover completely. Pat and press to release any air pockets and create a relatively uniform shape. Tuck the open ends neatly underneath. Wrap snugly in several layers of cheesecloth and secure each end tightly with butcher's twine.

Return the pan holding the poaching liquid to the hob and heat to 70°C. Carefully slip the galantine into the liquid and poach gently over a very low heat, turning the galantine occasionally and spooning the liquid over it to ensure even cooking, for about 1¼ hours until a thermometer inserted into the centre registers 63°C.

Remove the pan from the hob and leave the galantine to cool at room temperature in its poaching liquid for about 30 minutes. Place a plate on top to keep the galantine submerged in the liquid, then refrigerate the pan overnight.

The following day, remove the galantine from the pan and unwrap it from the cheesecloth.

Slice into rounds 1cm thick to serve. Slice as much as you plan to serve, then wrap the remainder well and store in the fridge for up to 4 days.

COU FARCI

This dish is a holiday speciality in the Gascony region of France and at the Fatted Calf. *Cou* is the French word for 'neck' and *farci* means 'stuffed'. In this recipe, the forcemeat of pork and veal is studded with braised chestnuts and chunks of cured foie gras, stuffed into goose or duck necks, and poached in a seasoned broth. **MAKES 4 STUFFED DUCK NECKS OR 2 STUFFED GOOSE NECKS; SERVES 8 AS A STARTER**

FORCEMEAT

600g boneless veal shoulder, cut into 2.5cm cubes
340g boneless pork picnic, cut into 2.5cm cubes
225g pork back fat, cut into 2.5cm cubes
1 teaspoon coriander seeds, toasted
1 teaspoon white peppercorns
3/4 teaspoon yellow mustard seeds
1 whole clove
1 allspice berry
3/4 teaspoon dried thyme
1/8 teaspoon ground mace
1/8 teaspoon freshly grated nutmeg 1 tablespoon fine sea salt

GARNISH

1 cup peeled chestnuts, fresh or frozen
500ml duck broth (see Basic Rich Broth, page 44)
60ml brandy
1 teaspoon fine sea salt
50g Foie Gras Torchon with Port and Quatre Épices (see page 275), finely diced

PANADE

180ml double cream
1 tablespoon brandy
15g fresh breadcrumbs

4 duck necks, or 2 goose necks

POACHING LIQUID

2 litres duck or chicken broth (see Basic Rich Broth, page 44)
60ml brandy
1 1/2 teaspoons fine sea salt

Place the veal, pork, and fat in a non-reactive bowl. In a spice grinder, combine the coriander seeds, peppercorns, mustard seeds, clove, allspice, and thyme and grind finely. Transfer to a bowl, add the salt, ground mace and nutmeg, and mix well. Mix with the meat, cover, and refrigerate overnight.

To assemble the garnish, in a pan, combine the chestnuts, broth, and brandy, bring to a simmer over a low heat, and cook for about 20 minutes, until the chestnuts are tender. Remove from the heat, season with salt, and refrigerate to chill.

Using a slotted spoon, transfer the chilled chestnuts to a cutting board, reserving the liquid, then quarter the chestnuts. Set aside with the liquid and foie gras.

To make the panade, in a small bowl, combine the cream, brandy, and breadcrumbs and mix well. Set aside.

Following the instructions for grinding on page 200, fit the grinder with the smallest plate and grind the meat once. Fold in the panade and grind again. Fold in the chestnuts and their cooking liquid and the foie gras and mix by hand for about 1–2 minutes, until the forcemeat holds together and the garnishes are evenly distributed. Cook a small sample of the mixture in a sauté pan, then chill the sample, taste for seasonings, and adjust if necessary.

Prepare the necks for stuffing. With the tip of a sharp knife, make a circle around the base of a neck, just above the wishbone, pressing all the way to the bone. Gently pull the skin completely away from the bone, turning the skin inside out. Trim away any blood vessels or glands. Using a trussing needle and butcher's twine, stitch closed the widest end of the neck. Knot the twine at both ends. Repeat with the remaining neck(s).

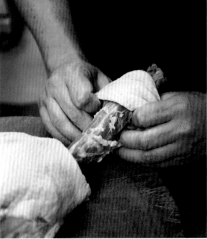

To make the poaching liquid, in a large pan, combine the broth, brandy, and salt, place on the hob, and heat to 70°C. While the liquid is heating, assemble a sausage stuffer and stuff the necks from the open end as you would salami (see page 232). Tie the open ends tightly closed with butcher's twine.

Transfer the stuffed necks to the poaching liquid and poach gently over a very low heat for about 25–30 minutes or until a thermometer inserted into the centre of a neck registers 60°C.

Take the pan off the heat and allow the necks to cool at room temperature in their poaching liquid for about 30 minutes. Place a plate on top of the necks to keep them submerged in the liquid, then refrigerate the pan overnight.

The following day, remove the necks from the pan and slice them into rounds 2.5cm thick to serve. Unsliced leftovers will keep in the fridge for 4–5 days.

Liver and Foie Gras Preparations

All foie gras is liver but not all liver is foie gras. Foie gras, literally 'fat liver' in French, is the result of *gavage*, a method of fattening a duck or goose in order to produce a very large, fatty liver with a meaty, silky texture and delicate, nutty flavour.

Raw foie gras is usually sold whole. Each liver consists of a large and a small lobe. The livers are graded A to C for quality. We highly recommend that you purchase only grade A foie gras, which is of superior quality, much easier to prepare, and only slightly more expensive than grade B. Grade C foie gras is of poor quality and usually only used in commercial preparations. Good foie gras will have an appealing colour that ranges from cream to pale peach, with little or no discoloration or blemishing.

In recent years, *gavage*, the process of feeding ducks and geese by hand for fattening their liver, has been at the centre of the foie gras controversy. It is important to decide for yourself whether or not you are comfortable with specific aspects of animal husbandry and to learn all of the facts before coming to a conclusion.

For most of their lives, ducks and geese that are destined to produce foie gras live on a farm just like their regular counterparts. In the last 2–3 weeks of their lives, they are fed by the method known as *gavage* two or three times each day. A small tube is inserted into the bird's throat and the bird is fed a mixture that usually contains some cooked corn. This practice prompts two questions: is it force-feeding and is it cruel? To answer these questions, you must keep a couple of things in mind. Ducks and geese are waterfowl and their throats, unlike ours, are designed to swallow whole, live, wriggling creatures. A food sac at the base of their throat allows them to store many meals' worth of food as they hunt. Their windpipe is located at the centre of their tongue, which eliminates any gag reflex during feeding. Gorging before migration is a natural part of waterfowl life, and they are adapted to store great quantities of fat in their livers without harming themselves. The fattened liver is healthy and normal, and if the birds were to return to regular feedings, their livers would return to their prefattened sizes.

Cases of animal cruelty have resulted from the production of foie gras, just as many cases of animal cruelty have occurred with all other types of farming. The geese and ducks are a valuable commodity and essential to the farmer's livelihoods. Stress or mistreatment will result in inferior foie gras and a loss for farmers. Responsible farmers care deeply about their charges and make the effort to give them good lives.

Foie gras is a treat, not something you eat every day. If you want to indulge now and again, get the good stuff.

The regular, unfattened livers of duck, goose, chicken and other poultry are also wonderfully flavoured and can be used to produce an array of charcuterie. Good-quality poultry produces good-quality livers, so always buy your livers from a farm you trust. Livers need to be perfectly fresh. Look for livers that are whole, with a deep red wine colour and no discernible odour.

Preparing Liver

Both regular livers and foie gras must be cleaned and trimmed before using. To prepare fresh duck or other poultry livers for cooking, rinse briefly in a colander under a gentle stream of cold running water. Pat dry with a paper towel and place on a cutting board. Use one hand to hold the liver in place and the other to pull any filaments of connective tissue or veins from it. Then, using a paring knife, trim away any discoloured areas or bile spots. Refrigerate the livers until ready to use.

To prepare foie gras for cooking, remove the liver from the fridge and allow it to temper at room temperature for about 1 hour. Foie gras is easiest to work with when it is just slightly cooler than room temperature. Using a paring knife, split the foie gras along the seam to separate the large and small lobes. Turn the lobes cut side up on your cutting board. Using the tip of the knife, expose the large

vein that runs from the top to the bottom, then slip the knife under the tip of the vein and gently tug it out, ideally in one piece. Long-nose pliers can be handy for this delicate operation. Using the same method, pull out any other visible blood vessels. Foie gras is fairly forgiving, and any usable foie gras displaced during cleaning can be gently prodded back into place.

DUCK LIVER MOUSSE WITH ARMAGNAC CREAM

'Sex on toast' was how one Fatted Calf regular described wedges of toasted *pan de mie* topped with this decadent duck liver mousse. In this desirable preparation, sautéed duck livers are enriched with butter and then exalted with Armagnac whipped cream to create a silky, sumptuous spread to be enjoyed on triangles of toast, in a Vietnamese-style *banh mi,* or, with the blinds drawn, au naturel, straight from the pot with a spoon. **MAKES ONE 1.5 LITRE TERRINE**

1kg prepared duck livers, cleaned and trimmed (see page 269)
1 tablespoon fine sea salt, plus a pinch
1/2 teaspoon freshly ground black pepper
1/4 teaspoon freshly ground white pepper
1/2 teaspoon curing salt no. 1 (optional)
2 tablespoons rendered duck fat
120ml duck gelée (see page 52), or 500ml duck broth (see Basic Rich Broth, page 44) reduced to 120ml then seasoned with 1 teaspoon fine sea salt
120ml Armagnac
340g unsalted butter, at room temperature
240ml double cream

Place the livers in a non-reactive bowl. Season with the 1 tablespoon salt, the black and white pepper, and the curing salt, if using, then cover and refrigerate overnight.

In a heavy sauté pan, melt the duck fat over a medium-high heat. When the fat begins to sizzle, add the livers and sauté for about 5 minutes. They should be rosy on the inside and yielding to the touch but not squishy. Turn out on to a platter to cool. Refrigerate, uncovered, for 2 hours.

Divide the cooked livers into three equal batches. Put one-third into a food processor and purée for 3 minutes. Slowly add one-third of the *gelée* and 1 tablespoon of Armagnac, followed by one-third of the butter. Process for another 2–3 minutes, until the mixture is very smooth. Using a rubber spatula, scrape the purée into a bowl. Repeat with the remaining two batches of livers, adding one-third of the *gelée*, 1 tablespoon of Armagnac, and one-third of the butter to each batch.

Stir the three batches together, then taste for seasoning and adjust if necessary. Set a fine-mesh sieve or tamis over a bowl, then pass the purée through it by scraping small amounts over the mesh surface with a spatula or plastic bench scraper.

In a bowl, using a whisk, whip the cream to soft peaks. Whisk the remaining Armagnac and a pinch of salt into the cream, then fold the cream into the liver purée and taste for seasoning.

If you will be serving the mousse *en terrine*, pack it into an earthenware pot. Alternatively, line a 1.5 litre terrine with cling film, allowing it to overhang the sides by at least 8cm, and fill with the mousse. Bring up the overhang to cover the top, then refrigerate for at least 3 hours to set. To turn out the mousse, remove it from the fridge and invert it on to a cutting board, then carefully lift off the terrine. Carefully peel away the cling film and slice to serve. The mousse will keep in the fridge for up to 4 days.

FOIE GRAS TERRINE WITH MADEIRA GELÉE

In this simple but stunning preparation, a whole lobe of foie gras is seasoned with *fleur de sel*, nestled into a terrine, and cooked until just set, then covered with a layer of Madeira *gelée*. The effect is a delight to both the eye and the palate. The glossy, caramel-coloured *gelée* surrounds the baked foie gras, seeping into the cracks on its surface and accentuating the opulent, nutty flavour of the liver. Sauternes or *vin santo* make an excellent alternative to the Madeira. **SERVES 8**

1 whole grade A foie gras

Fleur de sel

240ml duck gelée (see page 52) or highly seasoned gelatinous duck broth (see Basic Broth Making, step 12, page 43)

240ml Madeira

Hot brioche slices, to serve

Clean and trim the foie gras as directed on page 269. Lightly season the liver on all sides with the *fleur de sel*. Place in a non-reactive dish or container, cover, and refrigerate overnight.

In a pan, combine the *gelée* and Madeira, place over a medium heat and simmer for about 15 minutes, until reduced by half. Set aside to cool.

Preheat the oven to 140°C (gas1). Remove the foie gras from the fridge and leave to come to room temperature for about 30 minutes.

Bring a kettle filled with water to a simmer. Press the smaller lobe on the bottom of a small earthenware terrine or 1-litre soufflé dish, then mould the larger lobe around it. Fold a kitchen towel in half and place it in the bottom of a roasting tin. Set the loaded terrine or soufflé dish on the towel. Pour the simmering water into the roasting tin to reach a third of the way up the sides of the terrine. Carefully place in the oven and bake for 20–25 minutes, until a thermometer inserted into the centre of the foie gras registers 50°C. Remove from the oven and gently lift the terrine from the water bath.

Ladle off as much fat as possible without damaging the shape of the liver. Reserve the fat for another use. Allow the terrine to cool for 1 hour, then pour the reduced Madeira and *gelée* over the top. Refrigerate overnight to allow the flavours to marry.

To serve, let the terrine come to room temperature, then cut slices directly out of the dish and spread on to hot brioche. Serve with plenty of Champagne. Tightly wrap any leftover terrine in cling film and store in the fridge for up to 5 days.

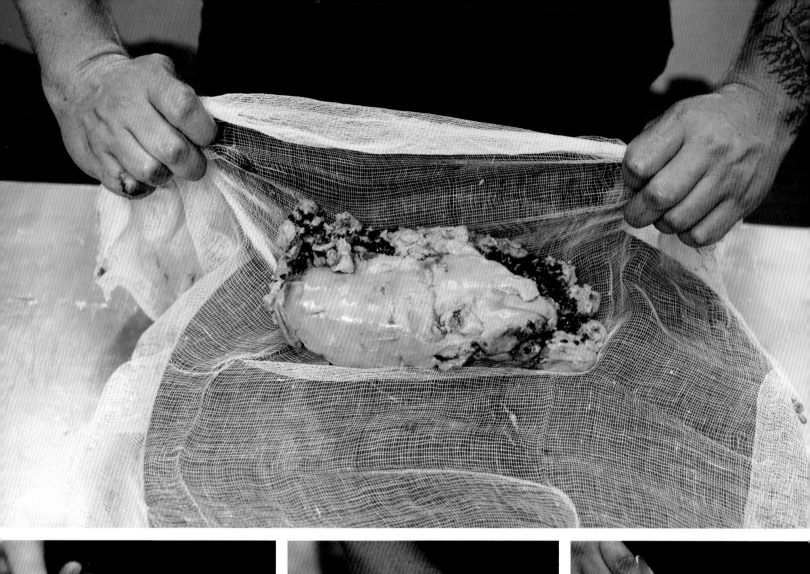

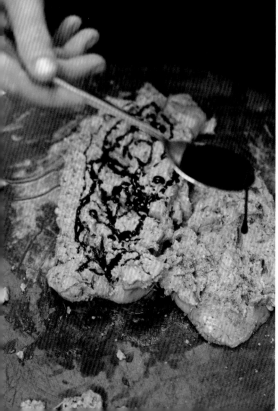

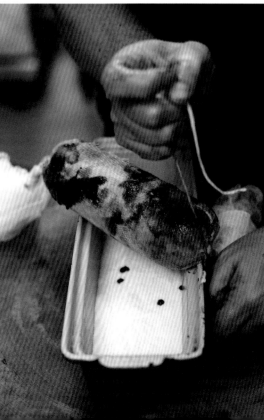

FOIE GRAS TORCHON WITH PORT AND QUATRE ÉPICES

Torchon is the French word for a 'rag' or 'dish towel', and a *foie gras au torchon* is a foie gras that is seasoned and then wrapped in a towel or cheesecloth to mould it prior to cooking. Some *torchons* are poached, but this one is buried in salt to 'cook'. The result is a dense, firm, buttery, well-seasoned foie gras with a fetching, dark inlay of reduced port and *quatre épices*. Slice the *torchon* and serve with warm brioche, shave with a vegetable peeler over a simple rocket salad, or top a steak hot off the grill with a thick coin of its buttery goodness. **SERVES 6–8**

1 whole grade A foie gras
Fleur de sel or Maldon sea salt
480ml port
1¹/₂ teaspoons quatre épices
2.5kg fine sea salt

Clean and prepare the foie gras according to the instructions on page 269. Lightly season the liver on all sides with *fleur de sel*. Place in a non-reactive dish or container, cover, and refrigerate overnight.

In a pan over a low heat, slowly cook the port for 20–30 minutes, until reduced to about 60ml. Pour into a small container, cover, and refrigerate overnight.

The following day, lay the foie gras on a cutting board or platter. Drizzle the reduced port over the cut side of each lobe, followed by a dusting of the *quatre épices*. Allow to marinate at room temperature for 1 hour.

Cut a double layer of cheesecloth about twice the length of the foie gras. Lay the large lobe lengthwise, cut side up, on the cheesecloth. Place the smaller lobe, cut side down, on top of the larger lobe. Fold the edge of the cheesecloth closest to you over the foie gras and squeeze to form a cylinder roughly 5cm in diameter. Foie gras is mostly fat and very malleable, so do not worry about damaging the liver. Roll the foie gras tightly in the cheesecloth and secure the open ends with twine. You should have a neat little parcel of relatively uniform dimensions that looks not unlike a salami.

Pour roughly one-third of the fine sea salt into the bottom of a deep container large enough to accommodate the *torchon* comfortably. Nestle the *torchon* on top of the salt, then cover completely with the remaining fine sea salt. The *torchon* should be buried with no portion of it visible above the top layer of salt. Cover and refrigerate for 3 days.

Gently unearth the *torchon* from the salt, brush off any salt that clings to the cheesecloth, snip the twine that secured the ends closed, and carefully unroll the foie gras. Cut a slice off the end to taste. It should be highly seasoned, rich, and very firm. If it seems a little bland, roll up the torchon in a fresh piece of doubled cheesecloth and bury it in the same salt for 12–24 hours longer.

Serve as suggested above. Tightly wrap any unused portion in cling film and store in the fridge for up to 1 week.

Braised Ham Hocks,
page 285

7

BRINED, CURED & SMOKED

Necessity is likely what drove us to discover the magic of dry curing, smoking, and brining in our quest for preserving meat for survival. But perhaps it is our human nature, our desire to control the elements, that led us to tinker with fire, water, air, and salty earth, turning necessity into the highly cultivated crafts of brining, dry curing, and smoking. These three finely tuned methods of preservation are often intertwined to create the salty, smoky, sweet, and spiced meats that traditionally tide us over during our lean times and accompany us on our journeys, providing sustenance and something more: flavour that we long for. Even today, when refrigeration, modern canning methods, and other technologies have rendered these ancient preservation methods obsolete, we hunger for bacon, have a hankering for a good pastrami sandwich, and cannot pass up a bite of salty sweet ham.

Brined

Think of a brine as a delicious bath that seasons, tenderises, and helps both to preserve meat and keep it juicy during cooking. It can be as simple as water mixed with salt or more complex, containing a host of aromatic ingredients. How long you brine and what goes into the brine depends on the type and size of the meat cut and the desired outcome. The meat is thoroughly submerged in this bath for anywhere from a few hours to achieve simple seasoning for smaller cuts, to a few days for larger cuts, or for up to several weeks when preservation is the goal.

Brining for Flavour

When you want to season meat thoroughly and retain moisture during roasting or smoking, brining is the way to go. It is especially beneficial for tougher or lean cuts, large cuts, or cuts on the bone. Through osmosis, brines penetrate the meat more effectively than regular salting, so that the interior of the meat is just as flavourful as the surface. Brined meats tend to cook at a slightly faster rate, and brining can offset the moisture loss that typically occurs during roasting, grilling, and smoking, keeping the meat moist. And if you happen to cook your brined roast a little more than you meant to, brining provides a cushion for variance, keeping your meat juicy and delicious.

Meats to Brine

Almost any meat can be brined, but pork and poultry are the most commonly brined meats. Both need to be cooked to a higher internal temperature than roasting or grilling beef, lamb, or goat cuts, and brining gives the meat added moisture to help it retain juiciness, even when cooked to higher temperatures. Very lean cuts, such as pork loin or chicken breast, or meats that will be smoked or slow roasted for a long time, such as spare ribs, benefit from brining for much the same reason. Unevenly shaped cuts, those comprised of several muscles, large cuts, or very bony ones such as a whole ham or turkey, also benefit from the ability of a brine to disperse seasonings thoroughly and evenly. Brining can even tenderise tougher cuts, such as leg or shoulder muscles.

Saltwater Science

There are two important processes that occur when meat is brined: osmosis and protein modification. Osmosis occurs when there is an uneven balance of solvent molecules on either side of a membrane: solvent redistributes itself and restores equilibrium by travelling from the area of higher concentration to the area of lower concentration. When meat is placed in a salty brine, osmosis is automatically triggered. Because the brine surrounding the meat has a higher concentration of salt than the liquid within the muscle, the muscle draws in the brine, allowing the seasonings to permeate the meat. But the salt that interacts with the proteins changes them. Cells plump and water that would normally flow out to create equilibrium becomes trapped. The cells both draw in and hold more water. The salt also acts on some of the proteins, breaking down their structures, loosening the connection between cells, and making traditionally tough cuts of meat more tender.

SALTY ENOUGH?

Prior to the standardisation of measuring devices, an egg was often used to test the salinity of a brine. If the egg was buoyant, the brine was good. If the egg sank, more salt was added.

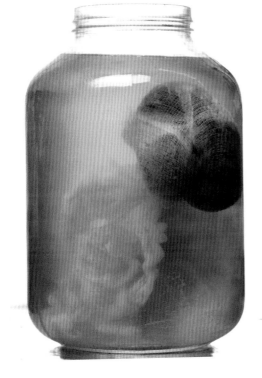

Brine Ingredients

The key to a brine is **salt**, and how much salt you use depends on the length of time the meat will be submerged in the brine. Fine sea salt is recommended as it dissolves easily in water. Always weigh the salt for a brine to ensure accuracy.

Sugar is used for flavouring and helps to balance the salt. It also provides a bonus: any sugar on the surface of the meat will caramelise during cooking, giving the meat a lovely golden sheen.

Aromatics such as spices, herbs, garlic, and onions, can be added to a brine to impart a variety of subtle flavours to meat. Be sure to add aromatics to the brine when it is still warm so that the flavours infuse thoroughly into the liquid.

Brining Basics

1. **Gather your equipment** and make sure you have ample space for refrigeration or have a very large cooler and plenty of ice. Choose a container that is made of hard food-grade plastic and about twice as large as the cut of meat you will be brining. You may also need a heavy plate or other weight to keep the meat submerged in the brine.

2. **Make the brine a day ahead** or earlier in the day so that it has plenty of time to cool to room temperature.

3. **Use very hot or boiling water** to dissolve the salt and sugar thoroughly in the brine.

4. **Add the aromatics while the brine is still hot** to facilitate the release of flavour. The aromatics can be tied in a cheesecloth sachet if you do not want to have to pick them off the meat after brining.

5. **Allow the brine to cool** to room temperature or just below before placing the meat in it. Never place meat into a hot or even warm brine.

6. **Fully submerge the meat.** Make sure that all of the meat is below the surface of the brine. Top with a heavy plate, if necessary, to ensure it remains submerged.

7. **Refrigerate the meat in the brine** immediately and keep it cold (at least 4°C).

8. **Label the brine container** with the contents, date, and time, so that you remember to pull the meat from the brine at exactly the right time. This is especially helpful if you are brining for several days or brining several cuts at once.

9. **Allow brined meat to air dry.** Once you remove the meat from the brine, place it on a tray or platter lined with a kitchen towel and allow it to air dry in the fridge. The surface must be completely dry before cooking so that the meat browns properly.

10. **Discard used brine.** Never reuse a brine.

ALL-PURPOSE POULTRY BRINE

This simple brine is good for seasoning everything from small game hens to big turkeys. Poultry is an especially good candidate for brining. The breasts and legs of most birds cook at different rates. Brining helps the leaner breast meat stay moist and juicy while the legs finish cooking. Be sure to allow the bird to dry fully before cooking to ensure crispy skin. The chart below lists the amount of brine necessary for birds of various sizes and the brining time for each. Depending on the size of your bird, you may need to divide or multiply this recipe to yield the required quantity of brine. **MAKES ABOUT 4 LITRES**

200g fine sea salt
100g sugar
4 litres boiling water

Place the salt and sugar in a large non-reactive container. Pour in the water and stir to dissolve the salt and sugar. Cover loosely and let cool for at least 4 hours or overnight, until cooled to room temperature.

Lightly rinse the poultry to be brined and pat dry. Submerge the poultry in the brine, top it with a plate to keep it submerged, and refrigerate for the desired brining time.

Line a tray or platter with a kitchen towel. Remove the poultry from the brine and discard the brine. Place the poultry on the prepared tray and refrigerate uncovered until the surface is thoroughly dry.

	WEIGHT	BRINE TIME	BRINE QUANTITY
Small birds	450–900g	6–12 hours	2–3 litres
Medium birds	900g–2.5kg	12–24 hours	4 litres
Large birds	2.5–5kg	24–36 hours	6–8 litres
Extra-large birds	4.5–9kg	36–48 hours	7–15 litres

GARLIC BRINE

This all-purpose pork brine can be used to season and tenderise shoulder, loin, spare rib, and other cuts. Make the brine the day before to allow the flavours of the garlic and spices to permeate the brine. The chart below lists the quantity of brine necessary for cuts of various sizes and the brining time for each. Depending on the size of your cut, you may need to divide or multiply this recipe to yield the required amount. **MAKES 4 LITRES**

20 cloves garlic, lightly crushed

2 tablespoons peppercorns

2 tablespoons yellow mustard seeds

2 teaspoons chilli flakes

2 teaspoons allspice berries

8 whole cloves

4 dried bay leaves

225g fine sea salt

270g sugar

4 litres boiling water

Place the garlic, peppercorns, mustard seeds, chilli flakes, allspice, cloves, and bay on a square of cheesecloth, bring the corners together, and tie securely with twine to make a sachet (or use a muslin bag). Place the salt and sugar in a large non-reactive container. Pour in the boiling water and stir to dissolve the salt and sugar. Add the sachet, cover the container loosely, and let cool for at least 4 hours or for up to overnight, until cooled to room temperature.

Lightly rinse the meat to be brined and pat dry. Submerge the meat in the brine, top it with a plate or other weight to keep it submerged, and refrigerate for the desired brining time.

Line a tray or platter with a kitchen towel. Remove the meat from the brine and discard the brine. Place the meat on the prepared tray and refrigerate uncovered until the surface is thoroughly dry.

	BRINE TIME	BRINE QUANTITY
Loin or pork chops	8–12 hours	2–3 litres
Spare ribs, loin back ribs, or belly	12–24 hours	4 litres
Whole loin, shoulder, or leg roast	48–72 hours	6–8 litres

CIDER-BRINED PORK CHOPS

Pork chops, which contain a cross section of both the loin and tenderloin, can be tricky to cook perfectly because the two muscles cook at slightly different rates. Brining this cut in effervescent hard apple cider ensures that both sides of the chop stay succulent and tender. This recipe calls for searing and then cooking the chops in the oven; however, they are also excellent grilled or smoked. If porterhouse chops are unavailable, pork rib chops (see page 105, step 4a) are a fine alternative.

SERVES 6

1 tablespoon black peppercorns

1 tablespoon yellow mustard seeds

3 whole cloves

5 allspice berries

3 dried bay leaves

250g fine sea salt

225g sugar

5 litres boiling water

180ml good-quality hard apple cider

6 bone-in pork chops

1 tablespoon lard

Place the peppercorns, mustard seeds, cloves, allspice, and bay on a square of cheesecloth, bring the corners together, and tie securely with twine to make a sachet (or use a muslin bag). Place the salt and sugar in a large non-reactive container. Pour in the boiling water and stir to dissolve the salt and sugar. Add the sachet, cover the container loosely, and let cool for at least 4 hours or for up to overnight, until cooled to room temperature.

When the brine has cooled to room temperature, stir in the cider. Lightly rinse the chops and pat them dry, then submerge them in the brine, topping them with a plate or other weight to keep them submerged, and refrigerate for 48 hours.

Line a tray or platter with a kitchen towel. Remove the chops from the brine and discard the brine. Place the chops on the prepared tray and refrigerate uncovered until the surface is thoroughly dry. (Once they are dry they can be wrapped in plastic and stored for 5 days in the fridge or for 6 weeks in the freezer.)

To cook, preheat the oven to 180°C (gas 4). Remove the chops from the fridge and bring them to room temperature for 20–30 minutes.

Heat a large, heavy ovenproof skillet over a medium heat. Add the lard and tilt the pan to coat the bottom evenly. Place the chops in the pan in an even, uncrowded layer. Brown them on one side for about 7 minutes. Turn the chops over, then slide the pan into the hot oven and cook for about 10 minutes more, until golden brown.

Serve hot.

Brining for Preservation

Brining meats is an age-old process of food preservation. Simple salt brines will inhibit the growth of bacteria, but they are not sufficient for long storage. When you want to use brining for preserving in addition to flavouring, you must add curing salt, or salt combined with sodium nitrate, to the brine (see chapter 1 for more on curing salts). Sodium nitrate kills the *Clostridium botulinum* spores that can cause deadly botulism along with other potentially harmful pathogens. Curing salt also gives brine-cured meats their characteristic pink colour. Without the addition of nitrates, these meats would turn a very unappealing shade of grey.

Brines with curing salt are made similarly to other brines except that the curing salt must be stirred into the brine after it has cooled to room temperature. It cannot be added to hot or warm brine because heat lessens its effectiveness. As an added measure of success, a brine pump can be employed. Basic brine pumps, which are available at speciality kitchen and butchery supply retailers and look like giant syringes, are used to inject the brine into thicker cuts or cuts on the bone to ensure it penetrates throughout. Injection also shortens the overall brining time. Injecting brine is not a substitute for submerging the meat in the brine, however. Instead, it should be done in conjunction with submersion.

SMOKED HAM HOCKS

In the cooler months, smoked ham hocks are a staple in the Fatted Calf kitchen. They make a flavourful addition to soups, beans, braised greens, or the Alsatian classic, *Choucroute Garni* (page 288). Shoulder hocks are generally preferred for this treatment as they have a bit more meat, but hock (hind) legs can be brined and smoked as well. **MAKES 4 SMOKED HOCKS**

1 tablespoon black peppercorns

1 tablespoon yellow mustard seeds

5 whole cloves

8 allspice berries

4 dried bay leaves

280g sugar

225g fine sea salt

4 litres boiling water

1 tablespoon curing salt no. 1

4 skin-on, bone-in hocks, about 340g each

To make the brine, place the peppercorns, mustard seeds, cloves, allspice, and bay on a square of cheesecloth, bring the corners together, and tie securely with twine to make a sachet (or use a muslin bag). Place the sugar and salt in a large non-reactive container. Pour in the boiling water and stir to dissolve the sugar and salt. Add the sachet, cover the container loosely, and let cool for at least 4 hours or for up to overnight, until cooled to room temperature.

Stir the curing salt into the cooled brine. Lightly rinse the pork hocks. Using a brine pump, inject about 60ml of the brine into each hock next to the bone. Submerge them in the remaining brine, top with a plate or other weight to keep them submerged, and refrigerate for 10 days.

Line a tray or platter with a kitchen towel. Remove the hocks from the brine and discard the brine. Place the hocks on the prepared tray and refrigerate uncovered overnight to dry thoroughly.

The following day, prepare your smokehouse following the guidelines listed on page 303. When the smoker reaches a temperature of 80°C, cook the shanks for about 3 hours, until a thermometer inserted into the thickest part of the meat not touching bone registers 65°C.

Remove the hocks from the smoker and let them cool to room temperature. Wrap tightly in cling film and refrigerate for up to 2 weeks or freeze for up to 3 months.

BRAISED HAM HOCKS

To unleash the mighty flavour and tender, succulent texture of brined and smoked ham hocks, they must first be braised. We find that most recipes that call for ham hocks fail to treat them to a lengthy simmer that releases their full potential. Many bean and soup recipes cook more rapidly than the hock, and although the hock will contribute flavour and a little fat to the pot, its meat will remain tough. The solution is to first braise the ham hock to tender perfection, which will yield both valuable braising liquor that can be used to enrich a pot of beans, braised greens, or soup and hock meat tender enough to be served as is or shredded off the bone. See photo on page 276. **MAKES 2 BRAISED HOCKS**

2 smoked ham hocks (see page 284)
2 litres water or pork, chicken, or duck
    broth (see Basic Rich Broth, page 44)
1 dried bay leaf
1 small onion, halved

In a pot, combine the hocks, water or stock, bay, and onion and place over a medium heat. Bring to a simmer, turn down the heat to low, and cook slowly just below a simmer. Cover or partially cover the pot if the liquid seems to be evaporating. Cook the hocks for about 3 hours, until the meat begins to pull away from the bone.

Cool and store the hocks in their cooking liquid until ready to use. The braised hocks will keep refrigerated for up to 4 days.

PICNIC HAM

This little cured and cooked 'ham' is not a ham at all: it's not a cut from the hind leg of the pig but rather from the shoulder or knuckle. This preparation is similar to that used for an Italian cooked ham, or *prosciutto cotto*: it is brined in white wine, coated with herbs, and gently steamed in the oven. Slice the ham into steaks for eggs benedict or use it to make a superb version of *Croque Monsieur* (opposite). **MAKES 2 PICNIC HAMS, ABOUT 450G EACH**

BRINE

1 teaspoon black peppercorns

1 teaspoon yellow mustard seeds

1 teaspoon allspice berries

3 whole cloves

1 dried bay leaf

200g fine sea salt

290g sugar

3 litres boiling water

1 tablespoon curing salt no. 1

360ml dry white wine

2 boneless pork shoulders,
   450–680g each

HERB RUB

1¹/₂ teaspoons freshly ground black pepper

15g finely chopped fresh flat-leaf parsley

15g finely chopped rosemary

15g finely chopped sage

2 tablespoons duck or pork gelée (see
   page 52) or broth (see Basic Rich Broth,
   page 44)

To make the brine, place the peppercorns, mustard seeds, allspice, cloves, and bay on a square of cheesecloth, bring the corners together, and tie securely with twine to make a sachet (or use a muslin bag). Place the sea salt and sugar in a large non-reactive container. Pour in the boiling water and stir to dissolve the salt and sugar. Add the sachet, cover the container loosely, and leave to cool for at least 4 hours or overnight, until room temperature.

Stir the curing salt and wine into the cooled brine. Using a brine pump, inject the centre of each shoulder with about 60ml of the brine, then submerge them in the remaining brine, top with a plate or other weight to keep them submerged, and refrigerate for 5 days.

After 5 days, remove the hams from the brine. Truss them tightly into relatively uniform cylinders and place them on a rack in a roasting tin.

To make the herb rub, in a small bowl combine the pepper, parsley, rosemary, sage, and *gelée* and mix well. Coat the hams evenly with the mixture and place on the rack.

Position an oven rack on the lowest rung in the oven and a second in the upper third of the oven. Preheat the oven to 110°C (gas ¼). Bring a small ovenproof pan filled with water to a boil on the hob and place it on the low rack.

Put the roasting tin on the upper rack of the oven and cook the hams for 45 minutes to 1 hour, until a thermometer inserted into the centre of each ham registers 60°C. Remove from the oven and leave to rest at room temperature for at least 10 minutes before slicing and serving warm.

If you will be eating the hams cold or at a later date, don't slice them. Let them cool to room temperature, then wrap them tightly in cling film and refrigerate for up to 10 days.

CROQUE MONSIEUR

This Parisian café classic may be the ultimate ham and cheese sandwich: crisp, buttery, piled with salty, sweet ham, and oozing with cheesy Mornay sauce. Make it with your own home-cured ham for a special treat. **SERVES 4**

8 slices *pain de mie* or other good quality
   sandwich bread
20g butter, melted
450g boneless picnic ham (see opposite),
   or other cured ham, thinly sliced
125g wholegrain mustard

MORNAY SAUCE
25g unsalted butter
1¹/₂ tablespoons plain flour
240ml whole milk
115g grated Comté or Gruyère cheese
55g grated Parmesan cheese
Pinch of freshly grated nutmeg
Pinch of ground cayenne pepper
Fine sea salt

To make the sauce, melt the butter in a small pan over a medium heat. Whisk in the flour until combined and then cook and whisk for 1 minute. Slowly whisk in the milk, turn down the heat to low and cook, stirring constantly, for about 5 minutes, until smooth and thick. Continue cooking and stirring for about 5 minutes more, allowing the sauce to thicken further, then remove from the heat and stir in the cheeses, nutmeg, and cayenne until the cheeses melt. Season with salt.

Preheat the oven to 230°C (gas 8). Lay the bread slices in a single layer on a baking sheet. Brush each slice with melted butter, then flip them over, buttered side down and spread the second side of each slice with mustard. Spoon 2 tablespoons of the Mornay sauce on top of each of 4 slices. Top the remaining 4 slices with a quarter of the ham, followed by 1 tablespoon of the Mornay.

Place the baking sheet in the oven for 8–10 minutes, until the sauce is bubbling and brown and the bottom of each bread slice is crisp and golden. Remove the pan from the oven and, using a spatula, marry each of the ham-topped slices with a Mornay-topped slice. Serve immediately.

CHOUCROUTE GARNI

This immensely hearty dish from France's borderland of Alsace is a classic of charcuterie cooking. Brined, cured, and smoked meats and sausages are nestled in a bed of tangy sauerkraut and braised in a combination of smoky hock broth and bright, crisp Alsatian Riesling until tender. Although brining and smoking all the meats yourself requires considerable planning and prep work, the results are immensely satisfying. Serve with boiled potatoes and plenty of strong mustard. **SERVES 8–10**

55g rendered duck fat or dripping

1 boneless pork shoulder roast, about 900g, brined in Garlic Brine (see page 281)

4 Belgian Beer Sausage links (see page 224) or similar sausage links

4 uncooked Cider-Brined Pork Chops (see page 282)

2 large onions, thinly sliced

480ml Alsatian Riesling or similar wine

2 Braised Ham Hocks with their braising liquid (see page 285)

3 litres sauerkraut, home-made (see page 319) or shop-bought

1 dried bay leaf

6–8 juniper berries

2 whole cloves

8 thick slices bacon, home-made (see page 299) or shop-bought

Preheat the oven to 150°C (gas 2).

First, brown the meats. Set a 9–10 litre sturdy braiser over a medium heat. Add the duck fat and heat until quite hot. Carefully place the brined pork shoulder in the pan and brown evenly on all sides. Transfer the roast to a large platter and set aside. Add the links and chops to the pan and brown evenly on both sides, then transfer them to the platter with the shoulder roast.

Lower the heat, add the onions to the pan, and cook, stirring occasionally, for about 8–10 minutes, until soft and translucent. Pour in the wine, bring to a simmer, and deglaze the pan, stirring with a wooden spoon to loosen the fond from the bottom. Add the hocks and their braising liquid, the sauerkraut, bay, juniper, and cloves. Distribute the pork roast, links, and chops evenly over the top, tucking them into the sauerkraut. They should be mostly covered, with just a bit peeking out from under the kraut. Lay the bacon slices across the top of the sauerkraut.

Cover the pan and place it in the oven. Cook slowly for 2 hours, then uncover and continue to cook for an additional 30–45 minutes, until golden brown.

Remove the pan from the oven. Transfer the pork shoulder roast and the links to a cutting board. Allow them to rest and cool for a few minutes. Slice each sausage on the diagonal into 4 or 5 thick slices, then tuck the slices back into the sauerkraut. Cut the pork roast into slices and lay the slices on top of the sauerkraut. Serve at the table, directly from the pan.

CORNED BEEF BRISKET

Corning is just an old-fashioned term for salting or brining. The spice-laden brine infuses streaky beef brisket with the pungent flavours of mustard, coriander, juniper, and bay in a traditional lengthy soak. The resulting corned beef can be simmered slowly in a traditional Irish American boiled supper, or rubbed with seasonings and smoked for the classic deli favourite, pastrami (see page 304). **MAKES 1 CORNED BRISKET, ABOUT 3.6KG**

900g fine sea salt

570g sugar

5 teaspoons black peppercorns

5 teaspoons yellow mustard seeds

2 teaspoons coriander seeds

3 juniper berries

10 whole cloves

3 bay leaves

11 litres boiling water

75g curing salt no. 1

1 fatty beef brisket, about 3.6kg

To make the brine, place the sea salt, sugar, peppercorns, mustard seeds, coriander seeds, juniper, cloves, and bay leaves in a large non-reactive container. Pour in the boiling water and stir to dissolve the salt and sugar. Cover the container loosely and leave to cool completely overnight.

The next day, stir the curing salt into the cooled brine. Lightly rinse the brisket and pat it dry. Using a brine pump, inject about 120ml of the brine into the centre of the brisket, then submerge it in the remaining brine, top it with a plate or other weight to keep it submerged, and refrigerate for 10 days.

Remove the finished corned beef brisket from the brine and rinse it well before cooking (see box, right). Discard the brine.

A whole brisket may seem like a lot of meat, but corned beef is the dish that keeps on giving. We always cook a corned beef twice as big as we think we might reasonably need in the hope that the next day we will be able to enjoy a hot corned beef sandwich heaped with sauerkraut, caramelised onions, and Muenster cheese. Any end bits can be shredded and sautéed with onions, cooked beets, and potatoes for a hearty breakfast of red flannel hash.

Choose a cooking pan that is wider and deeper than your hunk of beef is tall (you may need a deep roasting tin). Toss in a coarsely chopped leek, a couple of peeled whole carrots, a clove or two or three of garlic; a stalk of celery, and some sprigs of thyme and parsley. Set the corned beef brisket atop this aromatic bed and cover completely with cold water and perhaps a generous splash of beer. Place all of this in a preheated 150°C (gas 2) oven and simmer for 3–4 hours. Prod the brisket with a meat fork to ensure that it is tender and yielding before calling it quits. Let the meat rest in its cooking liquor for 20–30 minutes before carefully transferring it to a carving board and slicing it against the grain.

Seldom sticklers for tradition, we prefer to cook the required cabbage and potatoes separately – or sometimes not at all. (Sorry, Pop.) Boiled new potatoes with just a bit of butter and parsley will do, or a root vegetable mash is good, too. Far better than long-boiled cabbage is chopped cabbage sautéed with bacon until just tender and finished with a ladleful of the braising liquor. Creamy horseradish sauce and some strong mustard round out this fine Irish American feast – a supper so good that you may need to ration a bit for the next day's sandwiches before you sit down.

Dry Curing

Curing meat is a curiously addictive hobby. It evokes a sensation similar to watching a garden grow. You will need to do a bit of work at the outset, but once the initial labour is done, there is only tending, waiting, prodding, and anticipation until the aroma peaks and the muscle feels ripe to the touch, ready to harvest. You may experience a disappointment or two while you are learning the ropes, but once you get the knack, you might find yourself habitually smoking gorgeous golden slabs of bacon and stringing up spiced pork jowls. It feels as right and as satisfying to pull down a finished pancetta as it does to harvest plump pods of peas from the garden.

Dry curing is a simple technique. A generous layer of salt and other seasonings is rubbed on to the outside of a cut of meat to both flavour it and preserve it. Similar to brining, dry curing relies on osmosis to draw the salt and seasonings into the meat and protein modification to tenderise it. But unlike brining, water is not drawn in with the salt. Rather than adding moisture to the meat, dry curing dehydrates it, pulling out excess water, concentrating the flavours, and turning the meat a deep, rich colour.

Pork reigns supreme in the world of dry curing. Although dry-cured hams such as prosciutto are probably the best-known and most popular cured meats, nearly every part of the pig is a suitable candidate for dry curing, from trotters to loins. However, pork is not the only player. Cuts of duck, goose, beef, and even lamb can be dry cured.

Dry curing is usually one part of a two-part process. The initial dry cure begins when the meat is salted and moisture is allowed to drain off. This first part can take anywhere from a few days to a few weeks. Once the meat has lost the majority of its moisture, it can be either smoked or air-dried. Meats that are finished with smoke usually require less of an initial dry cure because smoking draws out additional moisture and provides an extra layer of preservation. Air-dried meats require a longer initial dry cure, after which they should be hung in a cool location to cure and firm up fully enough to eat without cooking. They can hang anywhere from a few weeks for smaller cuts, such as Guanciale (page 296), to a couple of years for a cured ham.

Ingredients for Dry Curing

Dry curing involves a familiar cast of characters.

Salt provides flavour, changes the texture by drawing out moisture, slows the growth of bacteria using dehydration, and regulates the fermentation process by lowering the water activity level. We use a fine sea salt for dry curing because it is most readily absorbed by the meat.

Curing salt is almost always used in dry curing because it keep meats from spoiling and, most importantly, it inhibits the growth of botulism. See page 9 for information on the two primary types of curing salt.

Sugar is sometimes used for flavouring and to help balance saltiness. Any number of sugars or sweeteners can be used, from simple cane sugar to maple syrup or honey. Dextrose is a popular choice for dry-cured meats because it is finely textured, easily absorbed, and has a less pronounced sweetness. Sugars also provide food for bacteria instrumental to the fermentation process.

Aromatics such as spices, dried herbs, garlic, and wine can also be used to impart a variety of subtle flavours to your cured meat.

Dry-Curing Basics

1. **Gather your equipment** and make sure you have ample space for refrigeration or another cool space to store your meat during the dry-curing process. Choose a container that is made of hard food-grade plastic or other non-reactive material. It should be at least a little larger than the cut of meat you will be curing. For some dry-cured items, you will also need a large plastic colander for draining during the initial cure. Restaurant supply stores often sell food-grade plastic pans or bins with perforated inserts that are perfect for dry curing and well worth the small investment.

2. **Prepare the dry-cure mix.** Carefully measure and combine the sea salt, curing salt, sugar, and aromatics, then mix well to distribute the ingredients evenly.

3. **Trim and weigh the meat to calculate the amount of dry-cure mix you will need.** Be sure to remove any silver skin or glands thoroughly and trim the meat into a somewhat neat or uniform shape before weighing it. The amount of dry-cure mix you will need is calculated according to the weight of the meat, and is generally about 3 per cent of the weight of the meat to be air-dried. This rule does not apply to meats that will be finished by hot smoking or cooking.

4. **Perforate the meat** by piercing its surface thoroughly and evenly with the tines of a sausage knife, the pointy end of a trussing needle, or a sharp skewer. The perforations allow the dry-cure mix to be more easily absorbed.

5. **Rub the meat** with the dry-cure mix, then massage the mix into the meat so that it penetrates thoroughly.

6. **Let the dry curing begin.** Place the rubbed meat in the chosen container. Label, date, and refrigerate the container. Drain off excess liquid and rotate as needed. The meat will lose most of its liquid in the first few days and then the process will slow.

7. **Finish the curing process by air-drying or smoking.** After the initial dry curing, most cured meats are finished either by smoking or by hanging for air-drying. Meats that are air-dried must be kept fairly cool, usually at about 10°C. A light mould will often form on the outside of the meat as it ages. In most cases, this is beneficial rather than harmful, creating a barrier that helps to protect the meat against potentially dangerous pathogens during the drying process.

PANCETTA ARROTOLATA

Pancetta is a type of cured pork belly, but unlike its sweet and smoky American cousin bacon (see page 299), Italian pancetta is air-dried after its initial cure. There are two basic types of pancetta, *tesa* and *arrotolata*. *Pancetta tesa* is flat, dried in its natural shape until it is so firm and dry that it can be eaten raw like prosciutto. *Pancetta arrotolata* is rolled, trussed and cured until just firm. Sliced or diced, it is the cured meat workhorse of the Italian kitchen, finding its way into tortellini and on to pizza, lending its delicately sweet porkiness to *sughi*, *insalate*, *zuppe*, and *verdure*, and barding lean meats such as pork tenderloin (page 172) with its ample fat, keeping them juicy and delicious during roasting. Start to finish, this pancetta will take a minimum of 4 weeks to prepare and cure. If you prefer a very firm pancetta, you can continue to let it hang for several more weeks. **MAKES ABOUT ONE 3-KG ROLLED PANCETTA**

1 skinless, boneless pork belly, about
    3.6kg or slightly larger

CURE MIX
1 tablespoon black peppercorns
2 teaspoons chilli flakes
1 dried bay leaf
5 allspice berries
85g fine sea salt
1 tablespoon dextrose
2 teaspoons curing salt no. 2
1/4 teaspoon freshly grated nutmeg

Trim and square off the pork belly so that it weighs about 3.6kg, give or take 30g. (Any trim can be saved for sausage or rillettes.) Lay the belly fat side down on a cutting board and pierce with the tines of a sausage knife, the pointy end of a trussing needle, or a sharp skewer, covering it completely with perforations roughly 1cm apart.

To make the cure mix, combine the peppercorns, chilli flakes, bay leaf, and allspice in a spice grinder and grind finely. Transfer to a small bowl, add the sea salt, dextrose, curing salt, and nutmeg, and mix well to distribute the ingredients evenly.

Place the meat in a non-reactive pan or container large enough to hold it flat. Rub the cure all over both sides and all of the edges of the belly, making sure to work the cure in between any crevices. When all of the cure has been massaged into the meat, lay the belly fat side down in the container, cover, and refrigerate overnight.

The following day, flip the belly over so the fat side faces up, then leave it to refrigerate over a second night. Turn it once again the following day, recover it, return it to the fridge, and leave it undisturbed to cure for another 12 days.

After 12 days, use a meat hook to hang the belly in the fridge (or other cold location) for 2 days to dry further. Place a pan or shallow bowl underneath the belly to catch any dripping juices.

When it is dry, remove the belly from the fridge and let it sit at room temperature for about 3 hours, until it is soft and malleable. Cover a large cutting board with cling film and lay the belly, meat side up, on top. Place another layer of cling film on top of the belly. Using a heavy meat tenderiser, rubber mallet, or rolling pin, pound the pancetta thoroughly, evenly, and without mercy for a full 5 minutes, until it is more pliable and easy to roll.

To roll and tie the pancetta, lay the belly, meaty side up, with its longest side closest to you. Fold the first 5cm of the side closest to you over on to itself and press down firmly. Continue to roll and press, making sure to eliminate any air gaps that can result in unwanted mould on the inside of the pancetta. When the belly is completely rolled, tie it as tightly as possible with butcher's twine as you would a roast (see pages 168–9), then rehang it in the fridge for another 2 weeks or until it is firm to your liking.

Pancetta keeps best when left whole or in large pieces. Slice off only what is to be cooked and tightly wrap the unused portion in cling film. It will keep well in the fridge for 3–4 weeks or in the freezer for up to 6 months.

GUANCIALE

Guancia is Italian for 'cheek', and *guanciale*, a speciality of central Italy, is cured pork jowl that has been rubbed with an aromatic blend of black pepper, hot chilli, and dried rosemary. Although it can be used in many of the same ways that pancetta is used, *guanciale* definitely has a personality all its own. When slowly crisped in a pan it gives off an intriguing, spicy, sharp fragrance that awakens the hunger. Sauté thin shards of *guanciale* with peas, asparagus, and spring onions, or toss crispy bits of sautéed *guanciale* with shredded cavalo nero, pecorino cheese, plumped currants and toasted pine nuts for a satisfying salad. Or, use it as the Romans do, for a robust bowl of *bucatini all'amatriciana* or *spaghetti alla carbonara.* **MAKES 1.5–2KG**

2.3kg skin-on pork jowls

CURE MIX
1 teaspoon black peppercorns
2 teaspoons chilli flakes
1 teaspoon dried oregano
¹/₂ teaspoon aniseeds, toasted
(see page 11)
1 dried bay leaf
55g fine sea salt
2 teaspoons curing salt no. 2

RUB
2 tablespoons coarsely chopped dried rosemary
3 tablespoons finely ground chilli flakes
28g freshly ground pepper

Trim the jowls of all glands. (The glands tend to reside on the inner meaty side of the jowls and are distinguishable by their off-white, shiny appearance.) Shave away any ragged edges so that the jowls have a somewhat uniform appearance. Prick the meat side of the jowls with the tines of a sausage knife, the pointy end of a trussing needle or a sharp skewer, covering it completely with perforations roughly 1cm apart. Leave the skin side undisturbed.

To make the cure mix, combine the peppercorns, chilli flakes, oregano, aniseeds and bay in a spice grinder and grind finely. Transfer to a small bowl, add the sea salt and curing salt, and mix well to distribute the ingredients evenly.

Rub the cure mix on to all sides of the jowls, place in a non-reactive container (it is fine if the jowls are stacked or touching), and refrigerate. Every day for 2 weeks, mix the jowls and rotate them from the bottom to the top to ensure that the cure is evenly penetrating each jowl.

After 2 weeks, remove the jowls from the fridge. Using a trussing needle

and butcher's twine, pierce the top of each one and run the twine through to create a loop. Hang them in a dry, cool location (10°C is optimal) to cure. Be sure to space the jowls at least 8cm apart to allow for proper airflow. After roughly 3 weeks, the jowls should feel fairly firm and can be taken down.

To make the rub, in a small bowl combine the rosemary, chilli flakes, and pepper and mix well. Massage the rub into the meat sides of each jowl. Wrap each one in plastic and refrigerate. Allow the rubbed jowls to sit overnight to allow the flavours of the rub to marry with the meat before using.

The *guanciale* can be sliced with its skin intact if you enjoy the chewy texture, or you can peel away the skin with a sharp knife before slicing. If you do peel off the skin, save it and toss it into your next pot of beans or minestrone to add complexity. Slice only what you need at the time, then tightly wrap the unused portion in cling film. It will keep well in the fridge for 4–6 weeks, or in the freezer for up to 6 months.

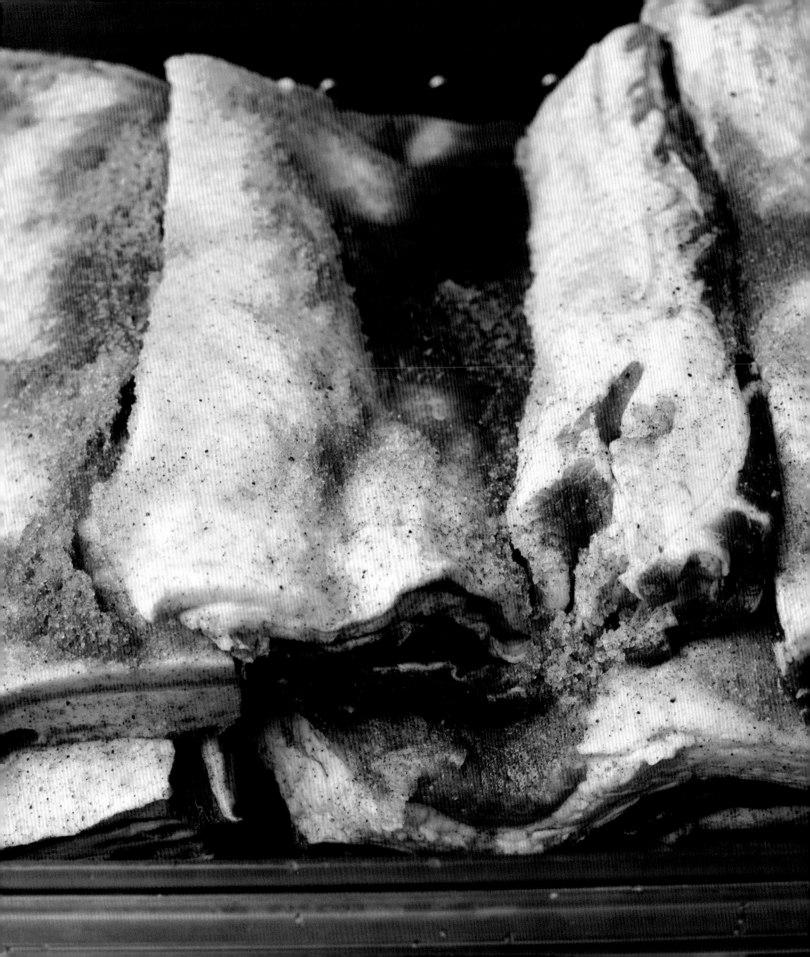

BROWN SUGAR–CURED BACON

A slab of bacon just out of the smoker is heaven on a platter. It is hard to resist tearing off a little end piece and shoving the still-steaming, salty, sweet meat into your mouth. Most commercially produced bacon is brined (or brine injected), but dry curing it – as we suggest in this recipe – produces meatier and more flavourful bacon that won't spit and curl when you cook it. Choose pork bellies that are fairly thick, with a fairly equal lean-to-fat ratio, keeping in mind that they will shrink slightly during the curing and smoking process. This a relatively quick dry-curing project, requiring just 3 days. **MAKES 2 SLABS, ABOUT 1.6KG EACH**

1 whole thick boneless, skinless pork belly, about 3.6kg

CURE MIX
450g brown sugar
340g fine sea salt
1 tablespoon curing salt no. 1
1 tablespoon ground cayenne pepper
1 tablespoon freshly ground black pepper

Use a sharp knife to square off the edges of the pork belly, then cut in half widthways to yield a pair of relatively square pieces of about the same size. Be sure to save any belly trim for sausage or rillettes, or other potted meats.

To make the cure mix, in a bowl, combine the brown sugar, sea salt, curing salt, cayenne, and black pepper.

Spread about 250g of the cure mix evenly over the bottom of a container large enough to accommodate the belly squares, one stacked on top of the other. Place a belly half, meat side down, directly on top of the cure. Rub an additional 250g of the cure on the edges and fat side of the belly, being sure to get the mix into all of the crevices. Place the second square of belly, meat side down, on top of the first. Rub another 250g of the mix on the edges and fat side of the belly, again being sure to get the mix into all of the crevices. Cover loosely and refrigerate overnight.

The following day, remove the belly slabs from the container and pour off any accumulated liquid. Switch the position of the slabs, placing the slab

previously on top on the bottom of the container and the bottom slab on top. Rub any remaining cure mix anywhere that it might be lacking. Recover and return to the fridge for another 24 hours. The following day, switch the slabs again.

After the belly slabs have cured for 3 full days, remove them from the pan and rinse them quickly under warm running water to remove any solids. Pat the slabs dry.

Prepare your smokehouse following the guidelines listed on page 303, and heat to 70°C–80°C. Place the belly slabs on a rack in the smoker and smoke for 2–3 hours, until a thermometer inserted into the centre of a slab registers 60°C.

Remove the bacon from the smoker and let it cool to room temperature, then refrigerate. The bacon must be well chilled before slicing.

Bacon keeps best in slab form. Slice only what you need at the time, then tightly wrap the unused portion in cling film. It will keep well in the fridge for 2–3 weeks or in the freezer for up to 6 months.

BRESAOLA

Bresaola is a northern Italian dry cured speciality made with lean beef eye of round. Dry curing tenderises this often tough cut and enhances its rich, beefy flavour. We prefer to case the eye of round in beef caps (see page 210) before hanging. Beef fat is more prone to spoilage than pork, and the cap protects the meat as it air-dries. It also gives the eye of round a neat and uniform shape, ensuring a more even dry cure. Thinly sliced, *bresaola* is a stunning deep red and has an earthy flavour that can be enjoyed either as part of an assorted antipasto or tossed into a salad of Italian leaves, persimmons, olives, and pecorino cheese. **MAKES ABOUT 1.5KG**

CURE MIX

1½ teaspoons black peppercorns

½ teaspoon coriander seeds

2 allspice berries

2 juniper berries

1 whole clove

50g fine sea salt

2 teaspoons curing salt no. 2

1.8-kg beef eye of round, completely trimmed of all external fat and silver skin

1 beef cap, rinsed and turned inside out (see page 210)

To make the cure mix, combine the peppercorns, coriander, allspice, juniper, and clove in a spice grinder and grind finely. Transfer to a small bowl, add the sea salt and curing salt, and mix well to distribute the ingredients evenly.

Check the eye of round carefully to make sure it is trimmed of all external fat and silver skin. Pierce the surface thoroughly and evenly with the tines of a sausage knife, the pointy end of a trussing needle, or a sharp skewer, covering it completely with perforations roughly 1cm apart.

Lay the eye of round in a large, flat non-reactive container (a large ceramic baking dish or food-grade plastic storage container works well). Rub the cure mix evenly over the surface of the beef, massaging it in well. Cover the container loosely and refrigerate overnight.

The next day, baste the beef with the liquid that has accumulated in the container. Place the eye of round in a large plastic colander or drain pan set over a container. Cover and refrigerate.

Every day for 2 weeks, turn the beef in the colander and discard any accumulated drippings.

After 2 weeks, case the eye of round in the beef cap. Remove the meat from the fridge. Drain off any excess water from the rinsed beef cap. Roll the open end of the beef cap on to one end of the eye of round, then pull it entirely over the meat, like a sock over a foot. Pull it taut and smooth out any air pockets. Using butcher's twine, and beginning at the closed end, tie the encased eye of round with butcher's twine as you would a roast (see pages 168–9), tying it as tightly as possible and pressing out any air pockets that may form as you go. Tie a knot and loop at the open end of the beef cap, as you would for salami (see pages 232–3). Hang the cased meat in a cool, dry location (10°C is optimal) for 6–8 weeks. It is ready when it feels firm, like a well-done roast.

To serve, peel back the casing on the portion to be sliced. Slice as thinly as possible into rounds, preferably using a meat slicer. *Bresaola* is best when freshly sliced and served right away, as it has a tendency to oxidise and dry out quicker than other cured meats.

Tightly wrap the unused portion in cling film. It will keep well in the fridge for 1–2 months.

Finishing with Smoke

What primeval yen draws us to the fire? The alluring perfume of wood smoke seems to beckon us outdoors to gather around the backyard barbecue and bear witness to the power of succulent smoked meats. The promise of a tender rack of spare ribs or a bite of beef brisket straight out of the smoker is enough to make you forget just how practical smoking really is.

Smoke has long been used as a method of meat preservation in cultures around the world. Wood smoke naturally slows the growth of harmful organisms. Phenolic compounds and formaldehyde contained in smoke are antimicrobial. Plus, smoke emits acids that create a protective layer on the outside of the meat that prevents the growth of surface mould and bacteria that can cause rancidity. Smoke also lowers the pH level on the surface of the meat, further discouraging the growth of harmful organisms. Used in conjunction with brining or dry curing, smoking has the ability to preserve meat for weeks, months, or even years.

The same phenols, carbonyl compounds, and organic acids that protect and preserve the meat also imbue it with an irresistible flavour. Refrigeration and other modern preservation methods have rendered smoking unnecessary, but the flavour of smoke has a depth and richness that is unparalleled. We continue to smoke our foods because we love the taste that it gives them. Smoking meat has become a craft that we practise for the pure joy it brings us.

The Smokehouse

Much ado is made about smokers, and there are some pretty expensive smokers on the market these days. But the smoker, at its most rudimentary, is really just a box with an internal or attached heat-smoke source and a chimney that helps to control the draft of smoke. It is the meat you put in it, the wood you use to smoke it, and the care you give it that make the finished product great. Choosing the smoker that is right for you and your needs is a personal decision that will depend on how often you use it, the general quantities of items you want to smoke, your available space and environment, and your budget. You may find that there are some bells and whistles that you must have and others that you can do without. That said, here are a few basic features to consider.

Control Yourself: A thermostat or dampers that allow you to control the draft is essential for controlling the temperature of the smoker. A thermometer to take the ambient temperature of the inside of the smoker and a probe thermometer to take the internal temperature of the meat are also necessary.

Hang Ups: You will need hooks and bars, racks or screens to hang or rest the meat on. Stainless steel is always a good choice for hardware, as it is sturdy, cleans up easily, and resists rusting.

Just Add Water: Regulating humidity inside the smoke chamber helps to keep the meat moist throughout. Cold smoking usually requires a

GAS STINKS

Mercaptans are the stinky organic compounds redolent of rotten eggs, decomposing cabbage, or bad breath that are added to propane and natural gas. They are highly detectable, even at only a few parts per million, which is exactly why they are used in odourless gas and propane. They immediately signal the possibility of a hazardous leak to the human nose. Although they serve a valuable function, food exposed to mercaptans can pick up unwanted bitter or unpleasant flavours. Stick to charcoal, electric, or all-wood-fired smokers to avoid the stink.

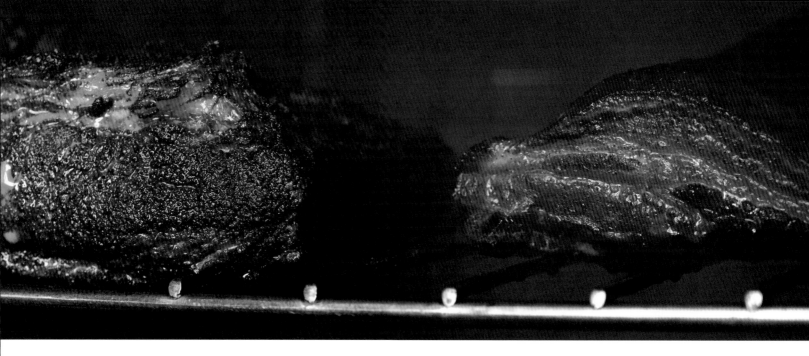

humidity level of about 75 per cent. The humidity level for hot smoking generally ranges from 50–60 per cent. Humidity sensors or meters are handy to help you to gauge the need for more or less humidity. To add humidity to the smoke chamber, use a water pan (some smokers come equipped with one, but you can also simply fill a pan with water and place it in the smoker). Soaking wood or wood chips for the fire also helps to create a more humid environment.

The Woodpile

All natural hardwoods and fruitwoods can infuse your meat with a range of different aromas. Each wood creates a unique smoke and flavour.

Alder is a reliable basic wood with a natural, subtle sweetness and delicate flavour that works well when blended with other, more nuanced woods, such as apple or cherry.

Almond is a clean-burning wood that can create a good bit of heat, making it excellent for starting a fire and creating a bed of coals to which you can add other aromatic woods.

Apple imparts a sweet and faintly fruity aroma that is perfect for bacon, ham, and pork chops but mild enough for poultry as well.

Cherry adds a light, sweet flavour when blended with stronger flavoured woods such as oak or hickory.

Grapevines have a distinctive flavour that works well with game birds. They are perfect for adding short bursts of heat or a quick infusion of smoke when grilling, but not great for longer smoke periods, as they typically burn quickly.

Hickory provides a strong and pungent smoky flavour that is classic for smoking beef but often too intense for more delicately flavoured meats.

Maple has a sweet and delicate aroma and is best combined with oak, apple, or alder for smoking hams, bacon, or birds.

Mesquite is oily, burns hot, and has an intense and sometimes overpowering flavour. It is best used in tandem with other types of wood.

Oak is the standard, all-purpose choice, heavily favoured by competition barbecue folk for its straightforward flavour. It can burn hot but, because it is very dense, it will also burn slowly, making it optimal for smoking large cuts.

Pecan is like a milder version of hickory. It burns slowly and at a lower temperature than most woods, but its fruity smoke is pungent and is best used in moderation.

Hot and Cold Smoking

There are two types of smoking: cold and hot. Cold smoking is used to flavour the meat rather than cook it. The meat is placed in an unheated chamber and the smoke is funnelled in from a firebox. Smokehouse temperatures for cold smoking should be below 38°C. Wood dust, chips, or pellets work best for cold smoking, as they will smoulder and smoke at a lower temperature. Hot smoking is akin to slow roasting with the addition of fragrant smoke and is used to cook the meat thoroughly. Smokehouse temperatures for hot smoking range from about 70°C to 85°C.

Smoking Basics

1. **Fat is your friend.** The natural layer of subcutaneous fat on the outside of the meat as well as the ample intramuscular fat or marbling will help to keep the meat moist and flavourful during its long, slow smoke.

2. **Form the pellicle.** Before your brined, dry-cured, or seasoned meats set foot in the smoker, they need to be air-dried so that they form a pellicle, a tacky coating of proteins on the surface of meat that allows smoke to better adhere to the surface. Air-dry the meats uncovered in the fridge, preferably on a rack or hanging from a hook to ensure they dry completely and evenly.

3. **Remove any lingering ash.** Make sure you remove the ash from the previous smoking session before you smoke.

4. **Gloves are good.** Heavy, non-flammable work gloves are great for protecting your digits when fussing with the fire and poking around in the smokehouse.

5. **Light your fire naturally.** Avoid using lighter fluid or any other chemical products that could produce unpleasant flavours.

6. **Let your coals burn down.** Before you place meat inside the smokehouse, let your coals burn down to the proper temperature.

7. **Start your meat fat side up.** As the fat begins to melt, it will baste the meat, keeping it juicy.

8. **Slow and low is the pit master's mantra.** There should be no flames, no flare-ups, and no sizzle at any point. If you do have a flare-up or if the smokehouse gets too hot, remove the meat and let things cool off before continuing. Smoking at too high a temperature will dry out your meat and reduce your yield.

9. **Use two thermometers.** You need one thermometer to monitor the ambient temperature of the smokehouse, and a second one to keep track of the internal temperature of the meat. Even if your smoked meats are well cooked and were made without curing salt, they will often have a rosy interior. Sometimes (if you're lucky) they will have a vivid pink ring around their exterior, the coveted smoke ring that is the Holy Grail of barbecue enthusiasts.

10. **Let it rest.** As with many cooked meats, let your smoked meat rest for at least 10–15 minutes before slicing and serving.

PASTRAMI

Pastrami is a project well worth the effort. One bite of this smoky, spicy brisket and you might never slum it at the supermarket again. Some traditional pastrami recipes call for steaming the meat in addition to smoking, but in this version the steaming and smoking happen simultaneously, resulting in a meltingly tender brisket, ready to slice and heap on to sturdy rye bread. **MAKES 1 WHOLE PASTRAMI, ABOUT 3.6KG**

Corned Beef Brisket (see page 290)

SPICE MIX
40g coriander seeds, toasted (see page 11)
35g peppercorns
1 tablespoon cumin seeds, toasted (see page 11)
40g yellow mustard seeds
55g Spanish unsmoked paprika
1¹/₂ teaspoons ground cayenne pepper
55g firmly packed brown sugar

Pat the brisket with paper towels to remove any excess moisture and allow it to air-dry at room temperature for about 1 hour. Place the brisket on a large baking sheet.

To make the spice mix, combine the coriander, peppercorns, cumin, and mustard seeds in a spice grinder. Transfer to a small bowl, add the paprika, cayenne pepper and sugar, and mix well to distribute the ingredients evenly.

Cover the brisket evenly with the spice mix, patting it on rather thickly. Refrigerate uncovered overnight.

The following day, prepare your smokehouse following the guidelines listed on page 303, and heat to about 75°C. If your smoker is equipped with a water pan, fill it. If not, fill a Dutch oven with water, bring to a rolling boil on the hob, and then carefully place it in the bottom of the cabinet or directly in the coals, depending on the design of your smoker. This regulates the temperature and helps to create steam to keep the brisket moist and tender. Place the brisket in the smoker. Tend to the fire and maintain a consistent temperature as needed.

Smoke the brisket for about 3 hours, until a thermometer inserted into the centre registers about 65°C.

Let the brisket rest for at least 30 minutes before slicing and serving, or let cool completely to room temperature before wrapping tightly in cling film and refrigerating if you prefer to enjoy it cold. It will keep well in the fridge for up to 2 weeks.

PULLED PORK

We can think of few better ways to spend a lazy summer day than to slowly smoke a pork shoulder to tender perfection and then shred it bit by bit into a steaming, irresistible mess. And there are few better ways to feed a crowd than with a mess of pulled pork.

Traditionalists and keepers of the barbecue flame need to know that this recipe does not claim to ascribe to a particular style, wear any pedigree of authenticity, or challenge the cherished recipe of your grandpa. But it is quite tasty and worth a go even for the seasoned pitmaster. The brine begins tenderising the meat and the long, leisurely smoke finishes the job. When all is said and done, this shoulder should be so tender it practically shreds itself. **MAKES A MESS OF PULLED PORK, ENOUGH TO SERVE 10–12**

1 skinless, boneless pork shoulder, butterflied and soaked in Garlic Brine (see page 281) for 3 days
Black Coffee and Bourbon Barbecue Sauce (see page 321)

Line a baking sheet with a kitchen towel. Remove the pork from the brine, place it on the prepared baking sheet, and refrigerate uncovered overnight to dry.

The following day, prepare your smokehouse following the guidelines listed on page 303, and heat to about to 80°C. Place the pork shoulder on a rack in the smoker. Open the shoulder up fully to expose a maximum surface area. Put a pan underneath the rack to catch any drippings that accumulate as the meat cooks. Tend to the fire and maintain a consistent temperature as needed.

Slowly smoke the pork for about 4 hours (far beyond well done), until a thermometer inserted into the thickest part of the meat registers 80°C. The connective tissue will have almost collapsed and the meat will be tender and obscenely flavourful.

Remove the pork from the smoker and set it on a tray to rest. When it is cool enough to handle, break it apart into chunks about the size of your thumb. Place the meat in a pan and stir in enough of the barbecue sauce and drippings to coat it well.

Place over a very low heat and keep at a bare simmer for 30 minutes, until the meat absorbs the sauce. Taste for seasoning and add more sauce as desired, then serve.

Cowboy Beans, page 325

8

ACCOUTREMENTS

One cannot subsist on meat alone. At the Fatted Calf we offer all manner of seasonal pickles and preserves, mustards and chutneys, crunchy vegetables and hearty legumes to accompany our roasts, pâtés, salami, and smoked and cured meats. Served atop a sandwich, alongside a sausage or spooned over grilled meats, these accoutrements serve to provide balance to the rich flavours of charcuterie.

BREAD AND BUTTER PICKLES

At the Fatted Calf we make bucketsful of these pickles throughout cucumber season. A little less sweet and a little more piquant than traditional bread and butter pickles, they are a great addition to burgers, an excellent accompaniment to smoked meats, and a welcome component of any charcuterie spread. You will find endless excuses to eat them with meaty goods or all by their crunchy, sweetly spiced selves. A jar of these irresistible chips never lasts long, which is why this recipe makes three jars. Choose cucumbers that are firm, bright green, and no more than 12cm long. **MAKES ABOUT 3 LITRE JARS**

1.8kg pickling cucumbers

1.4 litres cider vinegar, plus more if needed

650g firmly packed dark brown sugar

70g fine sea salt

1¹/₂ tablespoons yellow mustard seeds

¹/₂ teaspoon allspice berries

1 teaspoon celery seeds

¹/₂ teaspoon ground turmeric

1 onion, diced

1 red sweet pepper, seeded and diced

6 cloves garlic, lightly crushed

3 dried cayenne peppers

To prepare the cucumbers, place them in a large container, add cold water to cover, and let soak for 30 minutes. Remove from the water, trim away any remnants from the blossom ends, and make sure the cucumbers are squeaky clean, then drain in a colander.

Using a mandolin or a sharp knife, slice the cucumbers into chips about 3mm thick and divide them evenly among 3 sterilised canning jars.

To make the brine, bring the vinegar to a rapid simmer in a pan over a high heat. Add the sugar and salt and stir to dissolve, then lower the heat to medium. Add the mustard seeds, allspice, celery seeds, turmeric, onion, sweet pepper, garlic, and peppers and

simmer for about 5 minutes to allow the flavours to mingle.

Carefully ladle the brine over the cucumbers, distributing the seasonings as evenly as possible. Make sure the cucumbers are completely submerged. Top with additional vinegar if necessary, allowing about 1cm headspace, then cover them tightly. After the jars have cooled enough to handle, turn them upside down and leave to cool completely, then place upright in the fridge.

Let the pickles sit in the fridge for a few days before eating to allow their flavours to meld. They will keep in the fridge for 4–6 weeks. If you prefer longer storage, process the lidded jars in a hot-water bath (see page 311).

Canning pickles and preserves guarantees a ready supply of your favourite accompaniments year-round. The most common, safest, and simplest method is water-bath canning. Just place your chutney, *mostarda*, or pickles in sterilised glass jars and submerge them in a pot of simmering water to cook, or 'process'. As the jars cook in the water, their contents create steam that presses against the lid of the jar and forces air out of the headspace. When the jars have finished processing and they are left to cool, the steam condenses into water, leaving a vacuum that tugs the lid inward, sealing the jar and creating the characteristic canning pop.

To process your home-made pickles and preserves you will need a pot fitted with a rack and glass canning jars with lids. Traditional canning pots have a specially fitted canning rack and are highly recommended. A jar lifter and a canning funnel are also handy.

1. Start clean. Make sure all your cooking implements and work surfaces are clean to avoid any contamination.
2. Check your jars for any nicks or cracks. Such imperfections can harbour bacteria, and any fractures in your jars are likely to rupture during the water-bath processing. If you are using glass jars with lids and screw-on bands, the jars and bands are reusable if they are in good condition, but the lids must be used only once for safety.
3. Sterilise your canning jars and lids either by running them through the dishwasher or by washing them with hot, soapy water and rinsing them with boiling water.
4. Set the jars aside in a warm location while you prepare your pickles, chutney, sauce, or *mostarda*.
5. Fill a canning pot a little more than half full with water and bring to a lively simmer.
6. Fill the jars, leaving a little space between the contents and the rim – this is the headspace – to allow for expansion during processing. About 1cm is standard.
7. Wipe the rims clean and seal the jars with their lids. Be careful not to overtighten them as air needs to be able to escape during processing.
8. Place the jars in a canning rack and lower them into the water. Make sure the tops of the jars are covered with water by at least 2.5cm.
9. Cover the pot and turn up the heat to bring the water to a boil.
10. Once the water is boiling, set a timer. We generally process 500ml jars for 20–25 minutes and litre jars for 30–35 minutes.
11. Turn off the heat and let the jars sit in the water bath for 5 minutes. Lay a dish towel on your countertop. Carefully remove the jars from the pot, place them on the towel, and leave them undisturbed for about 12 hours.
12. Check each jar to make sure it is properly sealed: if the lid is concave and does not give when pressed, the seal is good. If the lid moves up and down, the seal failed and the jar should be stored in the fridge.

Properly sealed jars can be stored in a cool cupboard for up to a year. Once open, store refrigerated.

CLASSIC CUCUMBER DILLS

Unlike most pickles, which are made with a vinegar-based brine, the sourness of these pickles is derived from the fermentation process. Fermentation can be unpredictable, though, and over the years we started adding a small amount of vinegar to act as a catalyst and regulator, producing a very consistent dill pickle. When picking your peck of pickling cucumbers, you'll need to be choosy. The quality and freshness of the cucumbers is of utmost importance in this recipe. Pick firm, relatively small (somewhere around 10cm long), evenly shaped pickling cucumbers without bruising or abrasions, preferably with the stem end intact. Grape leaves, rich in tannins, help to preserve the cucumbers' crunchy quality. **MAKES ABOUT 1 X 4 LITRE JAR**

2 litres hot water

70g fine sea salt

120ml white wine vinegar

1.8kg very fresh pickling cucumbers

$1/2$ teaspoon yellow mustard seeds

1 teaspoon black peppercorns

$1^{1}/2$ teaspoons coriander seeds

2 allspice berries

1 dried bay leaf

1 dried cayenne pepper

3 cloves garlic, lightly crushed

Several sprigs dill, or 1–2 dill flowers

4 fresh grape or horseradish leaves

Pour the water into a large pitcher or other vessel with a spout. Add the salt and stir to dissolve. Let cool to room temperature, then add the vinegar.

Meanwhile, prepare the cucumbers. In a large container, soak the cucumbers in cold water to cover for 30 minutes. Remove from the water and carefully check each one for bruising. Discard any that are not up to snuff. Trim away any remnants from the blossom ends and make sure the cucumbers are squeaky clean. Any unwanted elements can ruin an entire batch. Drain the cleaned cucumbers in a colander.

Measure the mustard seeds, peppercorns, coriander, and allspice into a small bowl and mix well. Break the bay leaf and cayenne pepper each into thirds. Have the garlic and dill ready.

Place a grape leaf in the bottom of a 4 litre jar or pot. Arrange a third of the cucumbers on top of the leaf. Add a third of the mixed spices, 1 garlic clove, 1 or 2 sprigs of dill, and 1 piece each of bay and cayenne. Follow with another grape leaf and another layer of cucumbers and seasonings, then repeat with the third leaf and a third layer. You want to distribute the seasonings evenly throughout, but don't stress

out about being too exact. Top with the remaining grape leaf. Pour the cooled brine over the cucumbers to cover. Weight the cucumbers so that they stay submerged in the brine. We generally use a small ceramic plate, sometimes topped with a small jar of water for additional weight, although some pots come equipped with fitted stones specially made for the job. Lid the jar or pot and store in a cool, preferably dark location.

Check the pickles after 4 or 5 days. A thin layer of white film may appear on top of the brine. This is a natural by-product of the fermentation and can be skimmed off. Using tongs, remove a pickle and taste. It should be mildly sour or half sour. If this is your preferred degree of sourness, you can refrigerate the pickles immediately to slow the fermentation. For stronger-flavoured, brinier pickles, continue to ferment at room temperature for an additional 4 or 5 days.

Keep the finished pickles refrigerated. Although they will keep for about 3 months, they are optimal during their first month. Over time they may become less crunchy, but will still be delicious.

PICKLED RED ONION RINGS

Perky pickled red onions are a happy addition to any plate. Strew them over a *birria* taco (see page 144), toss them into salads, lay them atop your Ugly Burger (see page 203) or tuck them into your favourite sandwich to add a pungent bite. This simple brine can also be used to pickle pearl onions, a great addition to cocktails. **MAKES ABOUT 1 X 1 LITRE JAR**

3 large red onions, sliced into rings about 3mm thick
480ml red wine vinegar
65g sugar
2 teaspoons fine sea salt
3 or 4 sprigs marjoram or thyme
3 allspice berries
1 fresh or dried chilli (optional)

Place the sliced onions in the jar.

In a small pan, combine the vinegar, sugar, and salt and bring to a simmer over a medium heat, stirring to dissolve the sugar. Add the marjoram, allspice and chilli and simmer for 2 minutes longer. Carefully pour the hot brine over the onions and let them cool to room temperature. Cover and refrigerate overnight before serving.

Although these onions will keep refrigerated for several months, they are at their best during the first 4 weeks.

From left: Loulou's Sweet-and-Savoury Fruits (page 316), Cherry Mostarda (page 320), Bread and Butter Pickles (page 310), Cherry Mostarda (page 320), Green Tomato Chutney (page 321), Classic Cucumber Dills (page 312), Marinated Olives (page 317), Pickled Red Onion Rings (page 313)

LOULOU'S SWEET-AND-SAVOURY FRUITS

We first began selling our charcuterie at the Saturday farmers' market in Berkeley, at a stand just next door to Casey Havre, proprietress of Loulou's Garden and maker of incredible jams, pickles, and preserves. This recipe was handed down from Casey's great-aunt Louise, who would prepare it with her own home-grown figs, which she'd dried in the cellar. As a treat, Aunt Louise would wrap a slice of salami around the sweet-and-savoury figs for young Casey.

We love this sweet-and-savoury concoction paired with salami and other cured or smoked meats. Once the fruits are finished, reduce any leftover brine to a syrupy consistency and drizzle over grilled pork or duck. **MAKES 2 X 500ML JARS**

360ml red wine vinegar
120ml balsamic vinegar
170g honey
150g dried figs
75g raisins
130g dried apricots
180g dried pear halves
4 chiles de árbol
3 or 4 cloves garlic, peeled but left whole
2 dried bay leaves
2 small cinnamon sticks, or 1 large cinnamon stick, broken in half

Combine the vinegars and honey in a pan. Place over a medium heat and bring to a simmer to dissolve the honey.

Loosely pack the figs, raisins, apricots, and pears into 2 clean 500ml jars, putting about a quarter of the fruit in each jar. Add half the chillies, the garlic, bay, and cinnamon. Add the remaining fruits to the jars, followed by the remaining seasonings, dividing them evenly and leaving 1cm headspace at the top. Slowly pour the hot vinegar syrup over the fruits, immersing them completely. Release any air bubbles by gently prodding the fruit with a chopstick.

Put the lids on the jars and refrigerate. Allow to macerate for at least 1 week before serving. The fruits will keep well refrigerated for up to 2 months. Alternatively, after screwing on the lids, process the jars in a hot-water bath for 20 minutes as instructed on page 311, then store in a cool cupboard for up to 1 year, refrigerating the jars once they have been opened.

MARINATED OLIVES

A few slices of *Saucisse Sec aux Herbes de Provence* (see page 237), a glass of *pastis*, and a bowl of these olives are the next best thing to a vacation on the French Riviera. Warming the aromatic garlic, spices, and citrus in the olive oil allows the flavours to bloom. Be sure to serve these olives at room temperature or slightly warmed. **MAKES ABOUT 400G**

400g mixed olives (such as picholine, tournante, Gaeta, niçoise, and Arbequina)
120ml olive oil
1 dried cayenne pepper
$1/2$ teaspoon fennel seeds, toasted (see page 11)
3 cloves garlic, lightly crushed
1 dried bay leaf
1 sprig thyme
Wide strips of zest of 1 lemon or orange
120ml dry white wine

Place the olives in a colander and rinse briefly under cool running water.

In a small pan, combine the olive oil, cayenne, fennel seeds, garlic, bay, thyme, and zest. Place over a low heat and heat gently for 5 minutes, then stir in the olives just to warm.

Turn off the heat and add the white wine. Let cool slightly before serving or refrigerate for later use.

Although these olives will keep well refrigerated, they are best eaten within 3 weeks.

ROOT VEGETABLE CHOWCHOW

Chowchow is a sweet-and-spicy mixed-vegetable pickle perfect for topping a sausage sandwich or for enjoying alongside smoked or grilled meats. Although typically made with a mix of summer vegetables such as cucumbers, green tomatoes, and sweet peppers, Chris Lohman, our resident country boy and pickle aficionado, developed this recipe to take advantage of sweet, crunchy root vegetables in the winter months. **MAKES 2 LITRE JARS**

120g peeled and julienned carrot

120g peeled and julienned turnip

120g peeled and julienned swede

120g peeled and julienned celeriac

120g peeled and julienned daikon

240g peeled and julienned golden beetroot

140g thinly sliced green cabbage

1 onion, thinly sliced

1 tablespoon fine sea salt

480ml cider vinegar, plus more if needed

240ml water

200g sugar

2 teaspoons chilli flakes

2 teaspoons whole yellow mustard seeds

1¹/₂ teaspoons ground yellow mustard seeds

1 teaspoon celery seeds

1 teaspoon ground turmeric

In a large colander, toss together the carrot, turnip, swede, celeriac, daikon, beetroot, cabbage, and onion. Add the salt and gently massage into the vegetables to encourage the release of water. Set the colander over a bowl, cover with cling film, and refrigerate overnight.

The following day, pack the vegetables into 2 clean jars. Discard any liquid that has accumulated in the bowl.

To make the brine, in a saucepan, combine the vinegar, water, sugar, chilli flakes, whole and ground mustard seeds, celery seeds, and turmeric. Place over a medium heat and bring to a rolling boil, then remove from the heat.

Carefully pour the brine into the jars. Make sure the vegetables are fully submerged. Top with additional vinegar if necessary. Put the lids on the jars tightly. After they have cooled enough to handle, turn them upside down and continue to cool completely to room temperature.

Place the jars upright in the fridge to store. It is best to let the chowchow sit for a few days before using to allow the flavours to develop. It will keep in the fridge for about 1 month.

TRADITIONAL SAUERKRAUT

Crunchy, tangy, naturally fermented sauerkraut is a charcuterie staple, providing a counterpoint to the smoky, fatty meats in many classic preparations, from Alsatian-style *Choucroute Garni* (see page 288) to a stacked pastrami sandwich.

Sauerkraut is simply cabbage mixed with salt, and making your own is easy and satisfying. A sturdy stone pot is the optimal fermentation vessel. Its thick stone walls help to keep the kraut at a consistent temperature. Some pots, such as the ones made by German manufacturer Harsch Gairtopf, are even outfitted with weights designed to keep the cabbage submerged in its brine and with a gutter around the rim that allows you to create an airtight water seal. But we have also made many a great batch of sauerkraut using a bucket, a plate, a water jug, and a loosely knotted trash bag.

This recipe calls for 4.5kg of cabbage, which may seem like a lot, but keep in mind, once salted, the cabbage loses volume dramatically. If you would like to make more or less, just be sure to use 3 per cent salt by weight for whatever amount of trimmed cabbage you choose to use. **MAKES ABOUT 4 LITRES**

4.5kg green cabbage, outer leaves removed, quartered through the stem end, and cored

150g fine sea salt

OPTIONAL SPICES

1/2 teaspoon yellow mustard seeds

1 teaspoon black peppercorns

1/2 teaspoon allspice berries

2 or 3 whole cloves

Lightly rinse the trimmed cabbage under cool running water. Slice widthways into ribbons about 3mm wide. Place the sliced cabbage in a large bowl or container and toss with the salt. To help begin the extraction of water, lightly knead the salt into the cabbage.

When the cabbage begins to release some liquid, after about 30 minutes, pack it into a clean ceramic pot or bucket, a handful at a time, pressing down after each addition. When all the cabbage and its juices have been added to the pot, weight it. If your pot does not come equipped with fitted weights, use a heavy ceramic plate topped with a jug of water for additional weight. Press down lightly on the weights. The brine should rise up at least 2.5cm above the surface of the cabbage. If the cabbage has not produced enough brine, you can add 240ml water mixed with 1/2 teaspoon fine sea salt to raise the level of the brine.

To keep out unwanted elements, cover your pot or, if your pot does not have a lid, loosely cover it with a large trash bag. Store in a cool, dark location: a basement or garage is best.

Check the pot every few days to make sure the brine level stays well above the surface of the cabbage. Skim off any white foam that appears on the surface. Sauerkraut can give off a strong, sometimes unpleasant odour while fermenting, which is not indicative of the final product. The fermentation generally takes about 4 weeks, but it can take less time in the warmer months or more during cooler times of the year.

Taste the sauerkraut after 2 weeks and monitor its progress. When it has achieved a level of sourness you enjoy, remove it from the pot, pack it into clean glass jars, and refrigerate to arrest the fermentation. It will keep well refrigerated for 3–4 months.

CHERRY MOSTARDA

Whenever we travel to Italy, we stock up on *olio essenziale di senape* (essential oil of mustard), the ingredient that gives *mostarda di frutta*, a traditional northern Italian accompaniment to boiled or roasted meats, its sharp zing. Essential oil of mustard is different from the mustard oil commonly sold in Indian groceries. It is concentrated, extremely potent, and potentially dangerous if used incorrectly. Outside northern Italy, this *olio* is uncommon and highly regulated. We were quite lucky on a recent trip to stumble into the Antica Erboristeria Romana, a shoebox of a shop somewhere near the eye of Rome's labyrinthine centre that is lined from floor to ceiling with curious cabinets and tiny spice-filled drawers. When the proprietor reappeared from the back room with three tiny bottles and many Italian words (and gestures) of warning, we were elated. *Mostarda* would be on the menu!

Even without the few coveted drops of essential oil of mustard, this *mostarda* is a sweetly piquant foil to earthy *salumi*, decadent *pâtés*, and roasted duck or pork. It also makes an impressive stuffing for pork loins and shoulders. Although cherries are a personal favourite in this recipe, you can substitute apricots, peaches, nectarines, or peeled and diced pears for a similarly zesty concoction. **MAKES ABOUT 2 X 500ML JARS**

700g cherries

300g sugar

75ml white wine vinegar

40g yellow mustard seeds, lightly toasted and ground (see page 11)

40g yellow mustard seeds, lightly toasted (see page 11) but kept whole

3/4 teaspoon sea salt

1/2 teaspoon freshly ground pepper

1 or 2 drops essential oil of mustard (optional)

Stem and pit the cherries, reserving the pits. Place the pits on a square of cheesecloth, bring together the corners, and tie securely with twine to make a sachet (or use a muslin bag).

In a pan over a medium heat, combine the sugar and vinegar and heat, stirring to dissolve the sugar. Cook for about 10 minutes, or until syrupy. Add the cherries and the sachet and cook for about 10 minutes, until the fruit begins to soften. Add the ground and whole mustard seeds, cook for 2 minutes more to thicken, then remove from the heat.

Stir in the salt and pepper and carefully administer the drops of mustard oil. Leave to cool in the pan at room temperature. Remove and discard the sachet. Taste for seasonings and adjust if necessary, then pour into clean 500ml jars and refrigerate.

The *mostarda* keeps well refrigerated for 2–3 weeks. For longer storage, process the jars in a hot-water bath for 20 minutes as instructed on page 311, then store in a cool cupboard for up to 1 year, refrigerating the jars once they have been opened.

GREEN TOMATO CHUTNEY

At the tail end of tomato season, there is almost always an abundance of green tomatoes that must be harvested before the first frost. After eating our fill of fried green tomato and bacon sandwiches, we turn the remainder into this delicately sweet and spicy chutney. Slather it on a meat loaf (see page 254) sandwich, serve it alongside an assortment of terrines, or stuff it inside a pork roast before smoking or roasting. **MAKES ABOUT 3 X 500ML JARS**

220g firmly packed dark brown sugar

240ml cider vinegar

320g diced onion

75g diced red pepper

2 dried cayenne peppers

2 teaspoons peeled and freshly grated ginger

1 tablespoon yellow mustard seeds

4 allspice berries

1kg green tomatoes, cored and cut into
    large cubes

Fine sea salt

In a large pan, combine the brown sugar and vinegar over a medium heat and bring to a simmer, stirring to dissolve the sugar. Simmer, stirring occasionally, for about 5 minutes, until thick and syrupy. Add the onion, pepper, cayenne, ginger, mustard seeds, and allspice and simmer for 5 minutes. Add the tomatoes, season with salt, and turn down the heat to low. Cook, stirring occasionally, for about 12–15 minutes, until the tomatoes are softened and the

mixture is thick and jammy. Taste and adjust the seasonings if necessary. Pour into clean jars, leave to cool, then cover and refrigerate.

This chutney keeps well refrigerated for 3–4 weeks. For longer storage, process the jars in a hot-water bath for 20 minutes as instructed on page 311, then store in a cool cupboard for up to 1 year, refrigerating the jars once they have been opened.

BLACK COFFEE AND BOURBON BARBECUE SAUCE

Strong black coffee and a long pour of bourbon give this barbecue sauce a little backbone. It is essential for Pulled Pork (see page 307) and is a great mop for any smoked pork, beef, or chicken. Use good-quality coffee and bourbon and be sure to stir in any smoky meat drippings you have on hand. **MAKES ABOUT 1 LITRE**

360ml red wine vinegar

360ml ketchup

250ml pork broth (see Basic Rich Broth,
    page 44)

110g firmly packed dark brown sugar

120ml freshly brewed strong coffee

60ml bourbon

Smoked-meat pan drippings, for flavouring
    (optional)

Fine sea salt

In a pan over a medium heat, simmer the vinegar until reduced by about half. Add the ketchup, broth, sugar, coffee, bourbon, and drippings, stir well, and simmer, stirring frequently, for about 30 minutes, until the flavours blend harmoniously. Remove from the heat and season with salt.

Serve right away, or let cool, cover, and refrigerate for up to 1 week.

CHILLI TOMATO SAUCE

This sweet and richly spiced sauce came into being one September afternoon due to an overabundance of ripe tomatoes and peppers from our garden. It has become more commonplace than ketchup at our table, a go-to condiment for tacos, lamb burgers, or grilled chicken. You can also stir it into a simple beef stew or simmer Lamb and Herb Meatballs (see page 206) in it.

MAKES ABOUT 3 X 500ML JARS

1kg tomatoes

450g red sweet peppers

115–170g hot red chillies (such as jalapeño or serrano)

1 large onion, quartered

6 cloves garlic, peeled but left whole

1 teaspoon cumin seeds, toasted (see page 11)

¹/₂ teaspoon peppercorns, toasted (see page 11)

6 allspice berries

1 whole clove

60ml olive oil

Fine sea salt

Preheat the grill to hot.

Place the tomatoes, peppers, chillies, onion, and garlic on a large baking sheet. Slip the vegetables under the grill and cook, turning as needed to colour evenly, for 5–7 minutes, until the skins of the tomatoes, peppers, and chillies blister. Remove from the grill and let the vegetables cool until they can be handled.

Peel and seed the tomatoes, peppers, and chillies. Working in batches, combine the tomatoes, peppers, chillies, onion, and garlic in a food processor and purée until smooth.

In a spice grinder, combine the cumin, peppercorns, allspice, and clove and grind finely.

Heat the olive oil in a large, sturdy pan over a medium heat. Carefully pour in the purée, then add the ground spices and season with salt. Stir well, turn down the heat to low, and simmer for about 30 minutes, until the sauce has thickened.

Pour the sauce into clean glass jars, let cool, then cover and refrigerate.

The sauce keeps well refrigerated for up to 1 week. For longer storage, process the jars in a hot-water bath for 20 minutes as instructed on page 311, then store in a cool cupboard for up to 1 year, refrigerating the jars once they have been opened.

HORSERADISH SALSA VERDE

Fresh horseradish root is gnarly, dull brown, and virtually odourless. But when fresh horseradish is peeled and grated, it produces a powerful organosulfur compound, akin to mustard oil, that can clear the sinuses and bring on tears. It also adds a delicious kick to sauces paired with roasted and grilled meats, such as this spicy version of *salsa verde*.

Choose horseradish roots that are firm and wrinkle-free. Peel only as much as you intend to grate, leaving the remainder intact. Freshly grated horseradish should be used immediately or stored covered in vinegar to avoid discoloration and bitterness. **MAKES ABOUT 500ML**

45g chopped fresh flat-leaf parsley
15g chopped fresh oregano
35g rinsed and chopped capers
1 tablespoon garlic pounded to a paste in a mortar with ¹/₄ teaspoon fine sea salt

2 tablespoons freshly grated horseradish
Grated zest and juice of 1 lemon
180ml extra-virgin olive oil
Fine sea salt

In a bowl, combine the parsley, oregano, capers, garlic, horseradish, and lemon zest and mix well. Stir in the lemon juice and olive oil, then season with salt. Salsa Verde is best served the same day.

CÉLERI RÉMOULADE

Céleri rémoulade is as ubiquitous to the traditional French charcuterie as coleslaw is to the classic American delicatessen. In this simple preparation, raw celeriac is shredded and dressed with an assertive mayonnaise tempered with crème fraîche. Once you have peeled and julienned the celeriac, toss it immediately with the lemon juice, as it discolours quickly. This de rigueur salad is perfect for any picnic lunch or charcuterie spread and makes a great side for sandwiches. **SERVES 6–8**

4 celeriac, peeled and julienned
Fine sea salt
120ml freshly squeezed lemon juice
1 egg
180ml olive oil
1 tablespoon garlic pounded to a paste
120ml crème fraîche
70g Dijon mustard
¹/₂ teaspoon freshly ground black pepper
30g chopped fresh flat-leaf parsley

Season the celeriac with salt, then toss with 2 tablespoons of the lemon juice. Transfer to a sieve placed over a bowl and leave to weep while you whisk together the dressing.

In a bowl, whisk the egg until blended. Very slowly whisk in the olive oil to emulsify. When the emulsion is thick, whisk in the garlic, the remaining lemon juice, the crème fraîche, the mustard, and the pepper.

Discard any liquid that has accumulated in the bowl beneath the sieve, then transfer the celeriac to the bowl. Pour over about half the dressing and mix by hand to coat. Continue to add dressing to taste. You many not need all of it. Fold in about two-thirds of the chopped parsley, then sprinkle the remaining parsley over the top just before serving. This salad is best prepared and served the same day.

SIMPLE BEANS

Beans, simply prepared, are a great accompaniment to roasted and smoked meats. At the Fatted Calf we prepare beans almost daily for salads, as a side dish to enjoy with our roasts, and for adding to soups. We also sell a tremendous amount of dried heirloom beans for people to cook at home with sausages, ham hocks, and bacon. Our methodology for cooking beans is fairly simple: you must begin with good beans, preferably a flavourful heirloom variety and not more than a year old. Beans that have been sitting on the store shelf or in your cupboard for too long will not taste as delicious and are more likely to cook unevenly, resulting in beans that are crumbly or hard at the centre. For perfect, creamy-textured beans every time, buy beans from a good source and follow this simple method. **MAKES ABOUT 1.5KG**

450g dried beans
1 tablespoon lard or pan drippings
170g minced onion
1 dried bay leaf
Fine sea salt

Rinse the beans well in several changes of water, then place in a large bowl, add water to cover by 8–10cm, and leave to soak for at least 4 hours or up to overnight. Do not soak beans for longer than 12 hours, as oversoaking can lead to uneven cooking.

Preheat the oven to 150°C (gas 2).

Choose a good, ovenproof pot. Clay pot enthusiasts will tell you that beans cooked in a clay bean pot are better, and we are inclined to agree. But if you do not have a clay pot, a sturdy enamelled cast-iron pot or heavy Dutch oven also works well. Ideally, the pot should be large enough so that once the beans and water are in it, it is roughly only half full to allow for expansion and good air circulation during cooking.

Set the pot on the hob over a low heat and melt the lard, then add the onion and gently sweat for about 10 minutes, until tender and translucent. Drain the beans and add them to the pot along with the bay leaf. Add enough water to cover by about 5cm, bring to a simmer, cover, and place in the oven.

Monitor the progress of the beans every 30 minutes for the first hour, then every 15 minutes after that. Different beans cooks at different rates. Some beans cook rapidly and will be done in less than 1 hour. Others will require 3 hours or longer. Most fall somewhere in between. If the level of the water begins to drop and the beans are in danger of being exposed, add more water. When the beans begin to soften but are not quite done, season with salt. Start with a small amount, about 1 1/2 teaspoons, then add more to taste. Continue to cook the beans until they are evenly cooked, with a creamy texture and tender skin.

The beans can be served just as they are, straight from the pot with a drizzle of good olive oil and a handful of freshly chopped herbs, or they can be used in one of the following recipes. To store the beans, let them cool in their cooking liquid and then refrigerate in the liquid for up to 3 days.

COWBOY BEANS

The tiny pink *pinquito* bean, native to the Central California's Santa Maria Valley, makes a mean pot of beans. Simmered with sweet peppers and hot chillies, these beans are a perfect side for grilled and smoked meats. Popular throughout the western United States and northern Mexico, they are at home with both barbecue and Mexican-inspired dishes. You can substitute good-quality pinto or other small pink beans if the *pinquitos* prove elusive. **SERVES 8**

450g dried Santa Maria pinquito beans,
    cooked as directed in Simple Beans
    (see page 324)
140g finely diced bacon, home-made
    (see page 299) or shop-bought
1 white onion, diced
1 red pepper, seeded and diced
1 tablespoon chopped garlic
1 or 2 dried chiles de árbol
Fine sea salt
250g tomato purée
350ml pork broth (see Basic Rich Broth,
    page 44)
1 sprig oregano
1 sprig epazote (optional)

Have the beans ready in their cooking liquid. Slowly render the bacon in a Dutch oven or clay pot over a medium heat. When the bacon releases some of its fat and begins to brown, add the onion, pepper, garlic, and chillies and stir well. Season lightly with salt and cook, stirring occasionally, for about 10 minutes, until the onion and pepper are tender. Add the tomato purée, cook for 5 minutes longer, then add the broth, oregano, and epazote, if using.

Drain the cooked beans, reserving the cooking liquid. Add the beans to the pot along with enough of the cooking liquid to cover the beans by about 1cm. Bring to a simmer and adjust the seasonings if necessary. Turn down the heat to low and simmer gently for about 45 minutes to marry the flavours. Alternatively, if the coals in your barbecue are glowing, the beans can be simmered over indirect heat where their flavour will benefit from the perfume of the wood smoke. Serve hot.

FAGIOLI ALL'UCCELLETTO

In Italian, an *uccelletto* is a 'little bird'. This charmingly named dish translates as 'beans cooked in the style of little birds'. Simmered in a herby tomato and red wine sauce with a hint of cured pork, it is one of the most requested recipes at the Fatted Calf, where we make it frequently to accompany roasted or braised meats. For a simple, hearty supper, add a few links of Sausage Confit (see page 53) just before topping the beans with breadcrumbs and baking. **SERVES 8**

450g dried borlotti, cranberry, or cannellini beans, cooked as directed in **Simple Beans** (see page 324)

140g finely diced pancetta or guanciale, home-made (see page 295 or 296) or shop-bought

150g finely diced onion

70g peeled and finely diced carrot

70g finely diced celery or fennel

2 tablespoons minced garlic

1 dried cayenne pepper

240ml dry red wine

375g tomato purée

360ml pork or duck broth (see **Basic Rich Broth**, page 44)

2 tablespoons chopped fresh savory

1 tablespoon chopped fresh sage

115g fresh breadcrumbs

2 tablespoons pan drippings or lard

Preheat the oven to 150°C (gas 2).

Have the beans ready in their cooking liquid. Slowly render the pancetta in a Dutch oven or clay pot over a low heat. When the pancetta releases some of its fat and begins to brown, add the onion, carrot, celery, garlic, and cayenne, and stir well. Season lightly with salt and cook, stirring occasionally, for about 12–15 minutes, until the onion, carrot, and celery are tender. Pour in the wine and tomato purée and simmer until reduced by about one-third, then add the broth, savory, and sage.

Drain the cooked beans, reserving their cooking liquid. Add the beans to the pot along with enough of the cooking liquid to cover the beans by about 1cm. Bring to a simmer and adjust the seasonings if necessary.

Melt the drippings in a small pan over a low heat. Add the breadcrumbs and stir to coat.

Spoon the breadcrumbs evenly over the beans. Transfer the pot, uncovered, to the oven and bake for about 1 hour, until the breadcrumbs are golden brown. Serve hot.

BUTCHER'S LINGO

Butterflied: Describes a cut of meat that is sliced nearly in half, parallel to the cutting board, leaving a small hinge on one side that allows the meat to be opened like a book (or the wings of a butterfly). This technique is often used to create a pocket for stuffings.

Dry Ageing: A process in which meat (usually beef) is hung in a cold, controlled environment, usually several weeks, to improve flavour and texture. Natural enzymes break down connective tissue and encourage moisture loss, which concentrates flavour.

Flintstone Chop: A thick beef or pork loin chop with the rib and part of the belly still attached.

Forcemeat: A mixture of chopped or ground raw meat and fat used to make sausage, salami, galantines, terrines, pâtés, loaves, or other preparations.

Free-range or Free-roaming: Term typically used to describe poultry that has been allowed access to the outdoors for part of its life but is not necessarily pastured or raised outdoors.

Frenched: Describes a bone-in cut of meat (generally a rack or chop) that has been trimmed to expose the bone for the sake of appearance only.

Glove-boned or semi-boneless: Describes quail or small birds that have had all of their bones, except for the tiny wing and leg bones, removed.

Grain Finished: The practice of feeding cattle and other ruminants grain (usually cheap corn or soya beans on larger factory farms or feedlots, but sometimes more natural grains, such as barley and rice, on smaller-scale operations) to improve the flavour and increase the fat content of the meat.

Grass Fed: Describes the traditional, slow-growing method of allowing ruminants (cattle, sheep, and goats) to graze on fresh pasture to benefit the health and quality of life of the animals and to produce meat that is more healthful and more natural.

Halal: Term used for food products that are prepared according to Islamic law and under Islamic authority. It prescribes a specific method of slaughter that calls for making a deep incision in the throat to cut the carotid artery, windpipe, and jugular vein and then bleeding the animal completely before processing.

Heritage: A breed of farm animal with distinctive characteristics not suited to intensive modern farming and therefore endangered or rare.

Hormone- and Antibiotic-free:
A difficult to verify claim made by
the producer that no hormones or
antibiotics have been administered
to the animal.

Kosher: Food products that are
prepared according to Jewish law
and under rabbinical supervision. It
prescribes that animals be slaughtered
by a single, precise cut across the
throat, which causes the animal to
bleed to death.

Natural: Describes a product containing
no artificial ingredient or added colour
that is only minimally processed
(or processed in a way that does not
fundamentally alter the product).

Organic: Term that indicates that
the food has been produced through
approved methods that integrate
cultural, biological, and mechanical
practices that foster recycling of
resources, promote ecological balance,
conserve biodiversity and avoid the use
of synthetic fertilisers, sewage sludge,
irradiation, and genetic engineering.

Pastured: Term applied to animals
that are raised mostly outdoors, where
they can roam freely in a natural
environment and graze on grasses or
forage for plants, insects, roots and
other foods. This practice ensures
a better life for the animals, limits
environmental pollution, and yields
meat and dairy that is more natural,
nutritious, and better tasting than meat
produced in factory farms.

Porterhouse Chop: A beef, pork, or lamb
chop that includes the top of the loin as
well as some of the tenderloin or fillet.
Some consider this chop the best of
both worlds.

Poussin: A young chicken, generally
4–6 weeks old.

Primal Cut: A large section of meat that
is separated from the carcass during
butchering and from which retail cuts
are fabricated.

Silver Skin: The thin layer of white or
silver-coloured connective tissue found
on the meat of beef, pork, and lamb.

Spatchcock: A butchering technique
that removes the backbone from poultry
or game in order to flatten the body in
preparation for grilling or roasting.

Spring Lamb: A milk-fed lamb, usually
between 3 and 5 months old, that was
born in late winter or early spring and
brought to market before July 1.

T-Bone: A beef, pork, or lamb chop
similar to a porterhouse but with a
smaller section of loin attached.

Wet Ageing: A misleading marketing
term coined for meat that is stored in
vacuum-sealed plastic bags (which is
the industry norm, not a technique)
rather than dry aged.

ACKNOWLEDGMENTS

Not unlike the craft of charcuterie, making this book was a process – one that required the help of many hands.

We'd like to express our sincere gratitude to all of the incredibly hard working people in the farming community who strive to grow good, clean food. Thanks to Liberty Duck, Mariquita Farm, and McCormack Ranch in Northern California; and to the farmers, ranchers, and processors in the Heritage Food network: the Good Farm, the Layz S Ranch, and the Fantasma family and their staff at Paradise Locker Meats. Special thanks to Tim Mueller and Trini Campbell of Riverdog Farm, who have not only provided us with a bounty of produce and meat over the years, but also were kind enough to let us photograph their inspiring farm. It's easier to make good charcuterie when you start with great raw ingredients.

In the world of culinary arts, there is very little that is new or original. Each generation teaches the one that follows it. We have had the honour of working with many wonderful mentors over the years. Thanks to all of the cooks, butchers, and charcutiers who let us work in their kitchens, willingly shared their knowledge, and helped to keep the craft of charcuterie alive.

This book would not have been possible without the extraordinary collaboration of a great many people. Huge thanks go to our agent, Katherine Cowles, for prodding, pushing, and providing sage advice. Thanks to Aaron Wehner, Emily Timberlake, and Betsy Stromberg of Ten Speed Press for lending this project their support, enthusiasm, and incredible talent. Thanks to Christine Wolheim, Alex Farnum, and their many assistants who, despite intense heat, long days, and an attack of the killer bees, managed to produce gorgeous food photography.

To the incredible team at the Fatted Calf, thank you for prepping, schlepping, and keeping things rolling right along. An especially big thanks to Bailie, the beating heart of the Fatted Calf.

Thanks to the boys of Green Valley Ranch and our neighbours at Rancho Estrella for allowing us to cook in your kitchens and trespass on your beautiful land.

To our family and friends, thank you for all of your support and understanding. We apologise for the many missed birthdays and unreturned phone calls and hope that you enjoy this book.

INDEX

Published by Square Peg 2014

10 9 8 7 6 5 4 3 2 1

First published in the United States in 2013 by Ten Speed Press, an imprint of the Crown Publishing Group,
a division of Random House, Inc., New York

First published in Great Britain in 2014 by
Square Peg
Random House, 20 Vauxhall Bridge Road,
London SW1V 2SA
www.vintage-books.co.uk

Addresses for companies within The Random House Group Limited can be found at:
www.randomhouse.co.uk/offices.htm
The Random House Group Limited Reg. No. 954009

A CIP catalogue record for this book is available from the British Library

ISBN 978 0 22 409883 0

The Random House Group Limited supports the Forest Stewardship Council®(FSC®), the leading
international forest-certification organisation. Our books carrying the FSC label are printed on FSC®-
certified paper. FSC is the only forest-certification scheme supported by the leading environmental
organisations, including Greenpeace. Our paper procurement policy can be found at
www.randomhouse.co.uk/environment

Printed and bound in China by C&C Offset Printing Co., Ltd.